RETHINKING INNOVATION AND DESIGN FOR EMERGING MARKETS

INSIDE THE RENAULT KWID PROJECT

RETHINKING INNOVATION AND DESIGN FOR EMERGING MARKETS

INSIDE THE RENAULT KWID PROJECT

CHRISTOPHE MIDLER
BERNARD JULLIEN
YANNICK LUNG

PREFACE BY CARLOS GHOSN

CRC Press is an imprint of the
Taylor & Francis Group, an informa business
AN AUERBACH BOOK

Cover image © Groupe Renault.

This book is translated and adapted from the French book *Innover à l'envers, repenser la stratégie et la conception dans un monde frugal.* Christophe Midler, Bernard Jullien, and Yannick Lung, Dunod, 2017. EAN13: 9782100759880.

CRC Press
Taylor & Francis Group
6000 Broken Sound Parkway NW, Suite 300
Boca Raton, FL 33487-2742

First issued in hardback 2019

First issued in paperback 2022

ISBN 13: 978-1-03-247682-7 (pbk)
ISBN 13: 978-1-138-03720-5 (hbk)

DOI: 10.1201/9781315177984

Visit the Taylor & Francis Web site at
http://www.taylorandfrancis.com

and the CRC Press Web site at
http://www.crcpress.com

Contents

About the Authors

Christophe Midler is CNRS Research Director at CRG I^3 (Management Research Center) and Professor at Polytechnique School. His research topics are innovation strategy, product development, project and innovation management. He has explored these topics in various industrial contexts, especially the automotive industry, but also construction, electronics, chemistry, pharmaceutics, and high-tech start-ups. He has published his work in such journals as *Research Policy, International Journal of Automotive Technology and Management, Project Management Journal,* and *International Journal of Project Management.* He regularly communicates at conferences such as the European Group of Organizational Studies, the International Product Development Management Conference, and the European Academy of Management. Some of his recent books are:

- *Managing and Working in Project Society – Institutional Challenges of Temporary Organizations* (Rolf A. Lundin, Niklas Arvidsson, Tim Brady, Eskil Ekstedt, Christophe Midler, and Jörg Sydow). Cambridge University Press, 2015. (2016 Project Management Institute Project Management Literature Award.)
- *Management de l'innovation et globalisation, enjeux et pratiques contemporains* (with Sihem Ben Mahmoud-Jouini and Florence Charue-Duboc). Préface de Stéphane Richard, Dunod, Paris, Mars 2015.
- *The Logan Epic; New Trajectories for Innovation* (with Bernard Jullien and Yannick Lung). Dunod, Paris, 2013.
- *Working on Innovation* (Christophe Midler, Guy Minguet, and Monique Vervaeke). Routledge, New York, London, 2010.

Bernard Jullien is Assistant Professor at the University of Bordeaux, specializing in industrial economics. Since 2006, he has been Director of GERPISA, an international network for social science research specializing in the automotive industry and based in Cachan at the École Normale Supérieure. His research topics are at the intersection of industrial dynamics and sociopolitical analysis of public policies. In this perspective, he has co-edited two books with a political scientist (see below). About the car industry, he has dedicated the largest part of his research to studying motoring expenses, dynamics of consumption, the used car market and services. He has published his work in such journals as *Review of International Political Economy* and *International Journal of Automotive Technology and Management.* Among his books are:

- *Industries and Globalization, the Political Causality of Difference* (Bernard Jullien and Andy Smith, eds). Palgrave-MacMillan, 2008.
- *The Logan Epic; New Trajectories for Innovation* (with Yannick Lung and Christophe Midler). Dunod, Paris, 2013.
- *The EU's Government of Industries* (Bernard Jullien and Andy Smith, eds). Routledge, 2014.

Yannick Lung is Professor of Economics at the University of Bordeaux (France). His research interests are focused on the dynamics of technological and organizational change, mainly in the automobile industry, and their geographical dimensions. He has coordinated several international research programs on the car industry, being invited editor of scientific journals such as *International Journal of Automotive Technology and Management, International Journal of Urban and Regional Research,* and *European Urban and Regional Studies,* and has published numerous scientific articles and books. In English, he co-edited:

- *Cars, Carriers of Regionalism?* (with Jorge Carrillo and Rob van Tulder). Palgrave-Macmillan, London, 2004.
- *Ford 1903–2003: The European History* (with Hubert Bonin and Steven Tolliday). Editions PLAGE, Paris, 2003.
- *Coping with Variety. Flexible Productive Systems for Product Variety in the Auto Industry* (with Jean-Jacques Chanaron, Takahiro Fujimoto, and Dan Raff). Ashgate, Averbury, 1999.

He recently published:

- *Industrie automobile : la croisée des chemins* (with Bernard Jullien). La Documentation Française, Paris, 2011.
- *The Logan Epic: New Trajectories for Innovation* (with Bernard Jullien and Christophe Midler). Dunod, Paris, 2013.
- *Les trajectoires de l'innovation. Espaces et dynamiques de la complexité. XIX–XXIème siècles* (with Christophe Bouneau). Peter Lang, Bruxelles, 2014.

Foreword

This book tells the remarkable story of a project that is central to the global growth strategy of both Renault and the Renault-Nissan Alliance. The project drew its inspiration from Ratan Tata's initiative a decade ago to develop a low-cost car for India. Back then, I was the only one who believed that that was a good idea. There were many skeptics at the time, and their skepticism seemed justified when the Nano, following its launch in 2008, was less successful than expected. Yet I still believed this innovative concept was a promising one, with the potential to change people's lives in India and beyond.

As we all know, even the most brilliant idea is pointless if poorly executed. When I saw what became of the Nano, I wondered whether we could improve upon this promising idea. The Renault-Nissan Alliance had to take up the challenge, building on our own successful experience with the Dacia and Renault Logan, but I knew we would have to go further down the road of "frugal engineering" that we had taken with the Logan. At that time, the Entry occupied the mid-range segment, while the Duster held the high-end segment, particularly in India and in most developing country markets. The goal was to make a true entry-level car for these markets, which held immense potential.

Seven years later, we launched the Renault Kwid. It has been highly successful in India and, consequently, is much in demand in many other markets.

As always, when a project succeeds, many are eager to take the credit, but it is also essential to remember the challenges that were overcome along the way. This book meticulously analyzes the challenges with the Kwid project, providing its theme. Here, I would like to highlight three.

First, innovation depends on people and their skills. Think of a company without people. It is like imagining cuisine without the chef. Only people can make innovation a reality. We would not have had the Kwid if I had not had Gérard Detourbet to lead this project. Through his extensive automotive experience, his tenacity, and, especially, his success with the Entry program, he offered everything the project needed. At the time I offered him the challenge, I knew we were on to something, and that he would be able to lead his team and suppliers into new and uncharted territory.

Second, as the book illustrates, "frugal engineering" requires cutting-edge design expertise. For us, it involved combining the paramount need for frugality with the demands of a modern, solid, top-performing, high-quality car. One of this project's strengths was its leaders' knowledge of how to mobilize the significant engineering expertise found in Renault and Nissan by orienting its engineers toward targets very different from those to which they were accustomed.

Finally, on a strategic level, we had to foster innovation that went against all accepted norms. Had we approached Kwid in the traditional way, it merely would have led us to reproduce what our companies and competitors already knew. Faced with skepticism and resistance, a special system of governance had to be put in place to support this innovation and provide the necessary resources to ensure it bore fruit. I gave Gérard Detourbet the freedom he needed and, whenever required, I reiterated within Renault and Nissan that it was my project first and foremost, and that he could always count on me.

Managing innovation when it steps far beyond the boundaries of "business as usual" poses significant and diverse challenges. This book extensively analyzes those challenges and the tactics that can be utilized to help strengthen a company's collective capacity to innovate—a key issue for business today.

It also shows how these difficult issues can be turned into opportunities for companies that work hard to tackle them. Indeed, precisely because it was so difficult, the Kwid story cannot be easily and quickly replicated. The Alliance is ahead of the game in this strategic segment of entry-level vehicles for emerging countries, and it now holds a significant competitive advantage, considering the promise of tremendous growth these markets hold.

I would like to conclude with a comment on the CEO's role in the innovation challenge. Many analysts emphasize that innovation requires time and perseverance, but the average tenure of a CEO is becoming ever shorter. The Kwid story shows the importance of staying on course, even when storm clouds threaten.

We cannot look to short-term results if we want to encourage such approaches. The same holds true for the electric vehicle, another strategic bet I made 10 years ago. If the role of CEOs, as I believe, is to stimulate such original impulses and to support them so they stand out and develop, this means they must not be judged purely on their short-term prospects. This topic is ripe for future discussions in the field of management.

With that, I leave you in the capable hands of the authors, and I hope you enjoy the story of our adventure with the Kwid.

— Carlos Ghosn
Chairman and CEO
Renault-Nissan Alliance

Introduction

Innovation involves developing sophisticated products, usually incorporating new technologies, to achieve cutting-edge performance. However, it also increases costs. This is the type of innovation we see in industrialized countries: typically, it targets customers in the wealthiest areas (top of the pyramid) first, then rolls out through successive iterations in emerging or developing countries (trickle down)—what Clayton Christensen calls "sustaining" innovation.[1]

However, for several years, many multinational firms have been taking a different path: *reverse innovation.*[2] In place of conventional innovation strategies, reverse innovators adopt "trickle-up" strategies focusing on "the bottom of the pyramid"[3] to discover new compromises between the use value and cost of products. This strategy opens up new markets and customers that were previously neglected by extremely costly and complex innovations.

Adopting a competitive strategy for the high-growth markets of emerging countries means developing unique products that are focused on specific needs of emerging markets and are more likely to be produced locally. This is very different from the traditional approach, in which aging products, designed to be sold and manufactured in the home countries of major multinational firms, are just slightly adapted to emerging markets ("tropicalized").

---

[1] Christensen, Clayton M. *The Innovator's Dilemma: When New Technologies Cause Great Firms to Fail.* Boston, MA: Harvard Business School Press, 1997.

[2] Govindarajan, Vijay, and Chris Trimble. *Reverse Innovation: Create Far from Home, Win Everywhere.* Boston: Harvard Business Press, 2012.

[3] Prahalad, C. K. *The Fortune at the Bottom of the Pyramid: Eradicating Poverty Through Profits.* Upper Saddle River, NJ: Wharton School Publishing, 2005.

Successful reverse innovations are today critical for multinational companies. They also pose several problems:

- How to pursue reverse innovation alongside more conventional approaches in mature markets to generate revenue and build brand.
- What type of organization and design methodologies could be used to adapt products to local needs—particularly when R&D is rooted in mature western markets.
- How to roll out products in emerging markets where distribution networks are yet to be developed.

Most management books are long on mobilizing speeches and helpful suggestions, but short on the experiences of real-world executives who have actually tried to put such ideas into practice. With this book, we aim to answer the questions above by closely analyzing a high-profile case: the Kwid, a global car designed for emerging markets by Renault in partnership with Nissan.

Kwid hit the Indian market at the end of 2015 and did much to affirm the value of a reverse-innovation strategy for emerging markets. To some extent, the project aimed to repeat the successful strategy behind the Logan, the €5,000 car launched in Europe in 2004. However, it was far from being a Logan clone. At €3,500 (2,62 lakhs) as starting price, the Kwid rewrote the rules on pricing in a country where cars sold for between €2000 (1,5 lakhs), and €4000 (3 lakhs) are a major segment of the market. In terms of industrial ambition, it was based on a completely new engine and gearbox, whereas Logan had been built from existing components. While the Logan was initially intended to be a one-off, the Kwid developed a common platform for a range of Renault and Nissan products. And in terms of target markets, while Logan initially aimed to shift 60,000 units per year in Romania alone, the Kwid targeted hundreds of thousands of units over diverse markets.

The book is organized in two parts. The first traces the history of the Kwid project, from its genesis to its current commercial deployment in the Indian market. It is built in the form of a narrative. The second part reflects on this emblematic case from three perspectives:

1. Characterizing the very nature of such frugal breakthrough innovation, what we will call "fractal innovation," and the design process that could develop it.
2. Analyzing how such low-end strategies could emerge and deploy in large, established firms.
3. Characterizing the implementation of such reverse innovations within the global innovation footprint of multinational groups.

The book is based on research carried out throughout the Kwid project. We completed three research missions—in 2014, 2015, and 2016—giving us the opportunity to intensively interview key figures. This enabled us to follow the progress of operations, discover along with stakeholders the unexpected events they had to manage, and analyze the gap between ex-ante projections and ex-post realities. We also interviewed stakeholders who were not directly involved, but whose perspectives were important for understanding the project's progress. Finally, this survey was combined with an end-to-end analysis of internal project documents, allowing us to cross-reference stakeholders' accounts with the archives.

We would like to thank all the people who agreed to be part of this research, with commendable availability and transparency. Without their cooperation, these types of managerial experience would remain "black boxes" to outsiders. Instead, we can open them up and unfold a great deal of valuable management learning in the process. This study also received support from the Innovation Management Chair at École Polytechnique. We thank the sponsors of this chair (Air Liquide, MBDA, Renault, Safran, Seb, and Valeo), and also the Regional Council of Aquitaine, which supported the PROXIMO project.

And now, it is time to tell the story of the Kwid . . .

Part I

Incredible Kwid

Before we can tell the story of this project, we must draw a line to define its scope. When does it begin and end? How far should we go in analyzing developments and their impacts? In our case, we knew the beginning of the story relatively well, having written the Logan's story in 2013.[1] In a way, that project forms an important prologue to our story, at least for the key stakeholders who were involved in both projects. However, at the time of writing this book, the program had only reached the commercial start-up phase for Renault's Kwid product and Nissan's Redi-GO product in India, whereas this program has international ambitions. We focus only on the platform design (common to both companies) and the Renault product, partly because our knowledge of the Nissan side is incomplete and partly because the story of its vehicle is only just beginning.

This narrative is organized chronologically. It starts with the reconstruction of the emergence of the project in 2010–2011 to the decision to develop the program in its initial market in India. The second stage traces the difficulties of starting the program, difficulties that lead to the affirmation of an original development approach. We then analyze this process in terms of the organization of product-process development in the Indian location before detailing the supply relationships with local suppliers. After the design phase, we retrace the industrialization phase from the investment decisions to the product ramp-up and the commercial launch from October, 2015, to June, 2016. Finally, we outline the prospects for the international deployment of this new line of vehicles, especially in Brazil.

[1] Bernard, Jullien, Yannick Lung, and Christophe Midler. *The Logan Epic. A New Trajectory for Innovation.* Dunod, Paris, 2013.

Chapter One

Upstream Exploration: April 2010–October 2011

1.1 A Growth Strategy with Echoes of 1995

Upstream exploration for the project began in the first quarter of 2010. A working group was appointed in April under the leadership of Arnaud Deboeuf, who had been the Duster Program Manager and then the Entry Program Director.[1]

The strategic brief presented by Carlos Ghosn was very similar to the one used at the inception of the Logan project in Louis Schweitzer's strategic study of 1995.[2] As before, the CEO's objective was to increase the overall sales volume, the Renault-Nissan Alliance aiming to attain a 10% share of the global market. To achieve this ambitious objective, the Alliance had to target markets then growing, such as the "BRICs" (Brazil, Russia, India, China), and the segments with the largest sales volumes within these markets.

This strategic diagnosis spoke to the stability of the major trends in mobility from 1995 on, as well as the strategic stability of carmakers, which generally

[1] The Entry range has been derived from the Logan. In 2016, it had five different models: Logan (including the MCV and a pick-up), Sandero, Duster, Lodgy, and Dokker. This range is mainly sold under two brands: Dacia in Europe and some other markets, and Renault in most emerging countries (including Russia, India, and Latin America).

[2] Jullien, Bernard, Yannick Lung, and Christophe Midler. *The Logan Epic: New Trajectories for Innovation*. Paris: Dunod, 2013, p. 266.

developed growth strategies for emerging markets. However, the uniqueness of Renault's strategy lay in its "bottom-up" approach to specific products.

1.2 The Need for Coordination Within the Renault-Nissan Alliance

The second element of the exploration specifications was the need for a coordinated approach within the Renault-Nissan Alliance. This was a major change compared to the development of the Logan, which took place prior to Renault's acquisition of Nissan in 1999. The Logan project had emerged after Renault acquired Dacia as a niche project in Romania in September, 1999. At that time, the project was "sized" for 60,000 vehicles per year for the Romanian market, before being upsized to 150–180,000 for Eastern European markets, with the low investment making it possible to turn a profit even on such a low volume. However, by 2010, such an approach was out of the question—the aim was to go global from day one. Besides, initial studies would show that the strategy demanded a significant investment, which in turn implied large volumes. It was hard to envisage such scale without consolidating the sales of vehicle ranges that the two companies would each develop along the lines of the first expected model.

These phases of upstream exploration led to the study of multiple scenarios, as was the practice with large automotive groups. The two companies initially had different approaches, arising from their distinctive paths. The objective of the working group was to elaborate on their respective scenarios and converge on a shared view of the market and the two firms' product plans.

For Renault, the approach was comprehensive, but it emphasized the future of its strategy in India. There were several reasons for this. First was the crucial importance of this market, particularly in terms of its potential for the firm. Second was the fact that Renault already had a strong presence in Russia and Brazil with the Entry. Third was that the Chinese market, since the Yalta, which shared the global market between Renault and Nissan in 1999, fell largely within Nissan's sphere of influence, and was therefore less accessible for Renault.

By contrast, India in 2010 was a market Renault had already explored with the Logan, albeit with disappointing results.[3] To some extent, the Entry returned to the fore with the very successful launch of the Duster, in 2012. Unfortunately, it was a high-end product for India. The brand gained recognition in the market, but its presence was marginal. "With the Pulse, we re-styled the Micra for Renault. Renault wanted to enter the sub-four-meter market, which represents 40% of the market. However, 70% of the cars sold in the Indian market are below the starting price range of the group (see Figure 1.1).

[3] *Ibid.*

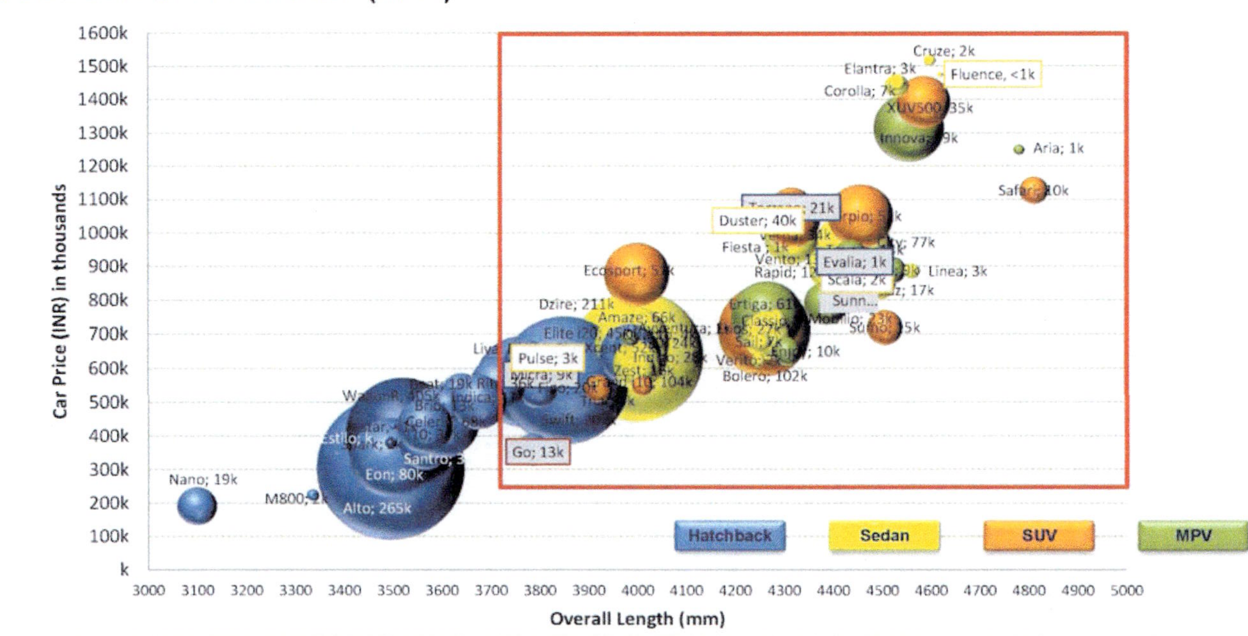

Figure 1.1 Analysis of the Indian market and Renault's position within it.

It was necessary to make a product below the Sandero in terms of the economic equation" (Patrick Le Charpy, Head of Advanced Design at LCI [see Section 1.3], who also heads the Indian design centers).

When it came to targeting the core Indian market, initial ideas centered on designing a simplified car based on existing platforms—Nissan's previous Micra, Renault's Logan—to limit investment (a "decontenting" strategy). However, the gap in the cost objective imposed by price constraints was too high: the sub-four-meter Indian market (beyond which taxation increased significantly[4]) required a selling price between 1.5 and 3.5 lakhs (€2,000–€4,500). This basically divided the cost objective by half in comparison with what had to be done for the Logan. And, just as with the Logan, the idea of a disruptive new product began to take root. Learning from the Entry program would play a key role, especially because the key figures of this exploration had experienced that program directly.

The explorations also examined alliances with partners. Among these were exchanges with Bajaj, the manufacturer of two- and three-wheel vehicles and the leader in rickshaws (small three-wheeled vehicles seen everywhere on Indian roads). *A priori*, this seemed an interesting approach, not least because of the frugality of this manufacturer. The idea explored in 2010 and 2011 was to develop a common platform and a model with the Indian company. Bajaj would commercialize it for B2B taxi companies, while Renault would be in charge of the private consumer market, which it would have to capture via an "automobile" network rather than via the distributors of two-wheeled vehicles, through which Bajaj reached households.

However, as this scenario was explored further, incompatible differences in approach soon surfaced. Bajaj developed its vehicle as a quadricycle, not a car. The alliance with Renault was advantageous, since Bajaj could integrate the European quadricycle standard in India, giving it a decisive competitive edge. For its part, however, Renault quickly saw that even if India met Bajaj's requirements, it would be impossible to recognize such a product as a car elsewhere in the world. Even in India, the scenario of creating a quadricycle standard seemed quite unlikely, given that competitor Tata would undoubtedly lobby hard against it. Tata had just launched the Nano, which met the Indian automobile standards. When the first images of the Bajaj prototypes were released, the writing was clearly on the wall: it would be impossible to put a Renault logo on such a product without seriously damaging the brand's image (see the Bajaj mockup in Figure 1.2). Thus the Bajaj hypothesis ended in June of 2011.

[4] A rule was introduced in India shortly after launching the Logan, which proved to be detrimental for competitive pricing. This vehicle was 25 cm longer than the standard of 4 meters.

Figure 1.2 Bajaj mockup. (© Renault, reprinted with permission.)

Nissan tackled the problem in a different way. In fact, its product planning department was developing a strategy to recreate a new entry-level brand, named Datsun, for cars below $7,000—somewhat similar to what Renault had done with Dacia for the Logan and the Entry range. Its recipe for covering the international market was to quickly develop a product plan comprising specific models in various countries, based on existing platforms. In the Indian market, the first product was the GO, derived from the earlier Micra and launched in 2014. In Russia, the DO, based on the Lada Kalinda platform, followed in 2015. However, it was clear that something other than the GO was needed in India. Datsun launched a pre-study with Defiance, an Indian engineering firm, to design a new platform *ex nihilo* (but a cheaper alternative to the GO) for a new product codenamed I2. However, after a few months, it was clear that this company's design capabilities were very limited for handling the end-to-end development of a new platform. The answer was a common platform supporting two distinct products: one Renault, one Datsun.

1.3 The Key Role of an Integrated Exploratory Approach: LCI

At Renault, the study advanced and took shape under the impetus of the Cooperative Innovation Laboratory (LCI) and the Design Department. LCI was a new entity, founded in 2006 by extending DISA (the Innovations and Vehicle Synthesis Department), the in-house vehicle engineering department established to explore new vehicle concepts by Yves Dubreil, former Director of the Twingo project and a key figure for promoting innovation within the company. LCI combined product, engineering, and design competencies, and exploring the Indian market would be an ideal opportunity for it to demonstrate its capabilities in contributing to new relevant answers using its new creative approach. Hence, it was mobilized intensively for this project.

"At the start of the project in 2010, the director for pre-projects, the LCI director, and I met over lunch at the Technocentre's canteen," recounts Patrick Le Charpy. The initial idea proposed by Philippe Klein, the Renault product plan director, was to use the Micra as the base, and LCI duly worked on this idea for six months. In the beginning, the Design department quickly constructed two small-scale mock-ups, with dimensions extrapolated from the existing vehicles—somewhat like a mini-Zoe and a mini-Captur. However, they failed to convince.

At the in-house Indian design studio in Mumbai, Ramesh Gound, one of the three designers on the team of Jean-Philippe Salar, worked on a small crossover brief, a "baby Duster" inspired by the Indian success of the Duster, which had just been launched in India. Gérard Detourbet, who saw the sketches in January, 2011, marked it immediately as the preferred path. The same month, during a project review at the in-house design office in Mumbai, Renault Design Director Laurens van den Acker, who joined the company in 2009, and Patrick Le Charpy discovered Ramesh Gound's "green" sketch and validated this choice (see Figure 1.3).

"We knew straight away we were on the right path," recalls Laurens van Den Acker. Indeed, LCI pursued this direction by mobilizing a team of 15 to 20 people, along with the product and architecture design departments, and no alternate scenarios were developed. With this design, LCI believed it had the key to completing the project and to becoming an important, possibly indispensable authority on stimulating new innovation within the group. So it worked toward this goal: "At LCI, 70% of our activity is based on requests and we make our own decisions for the remaining 30%. This was the case for this project. It was

Figure 1.3 The "green" sketch by Ramesh Gound, a member of the Renault Design Studio, Mumbai, January, 2011. (© Renault Design Studio, reprinted with permission.)

Figure 1.4 The converged mock-up, June, 2011. (© Renault Design Studio, reprinted with permission.)

important for LCI to show it could provide added value by proposing new innovations," says Patrick Le Charpy.

From January to July, 2011, LCI worked on the project. This work culminated in a so-called "converged" mock-up (see Figure 1.4) that incorporated economic and ergonomic constraints. "Our design was, as always, in competition with other more conventional approaches developed at the Technocentre. Tracks that favored rather the internal habitability against the strength and dynamism of the exterior design. But people who knew the Indian market clearly preferred our product vision," recalls Jean-Philippe Salar—a position that eventually prevailed.

This does not mean that the finalization of the design was easy. "There were conflicts over complying with the economic constraints. For width, Design wanted 1,600 mm, while Project wanted 1,500 mm; for the size of the wheels, Design wanted 14 inches and Project wanted 13 inches," said Laurens van den Aker. Ultimately, the project won: Gérard Detourbet, former program director of Entry, who worked single-mindedly on this project from July, 2011, but was involved in it right from the beginning, proposed 13-inch wheels, prompting Laurens van den Aker to note: "I have never been asked to use 13-inch wheels before . . ." The agreement was as follows: the vehicle was to have 13-inch wheels, but the wheel arch would accommodate 14 inches for markets in which it was mandatory (namely, Brazil). Moreover, Gérard Detourbet accepted that plastic rods for the wings would be provided, to increase the apparent visual size of the 13-inch wheels.

Figure 1.5 Stills from the film showing the converged mock-up XBA operating on the road, July, 2011. (© Renault Design Studio, reprinted with permission.)

In July, 2011, the converged mock-up was completed, and LCI made a brief film to showcase the product's suitability for Indian traffic (see Figure 1.5). This communication tool played a crucial role in prompting the decision to start the project. Arnaud Deboeuf presented it no fewer than three times to Carlos Ghosn before he finally gave the green light.

1.4 The Alliance Converges on a Strategic Scenario

Meanwhile, with the hypothesis of an all-new Renault car, now codenamed "XBA," becoming more detailed, an agreement had to be reached with Nissan. This entailed a complex period of converging product plans and engineering, finally accomplished by January, 2012, at which point the scenario of the program was finally frozen. The steps along the way to this goal were as follows.

- The Alliance meeting in October, 2011, adopted the strategy of a single platform for products initially intended for India and Indonesia, and produced in India by the Alliance plant. Other markets would be considered later for any additional deployments, with affordability assessed on a case-by-case basis.
- Two products, a Nissan product codenamed "I2" and a Renault product dubbed "XBA," would be produced on this platform (see Figure 1.6).

Figure 1.6 Renault Kwid (left) and Datsun redi-Go (right) contrasted final designs to differentiate the product identities. (© Renault, reprinted with permission.)

- At the organizational level, the Alliance agreed to entrust the development of the platform to a single common development entity located in Chennai, India.

It all boiled down to one question: should we develop a new small engine, as Renault proposed, or buy an existing one, as proposed by Nissan Product Planning Director Andrew Palmer, who was then in discussion with Mitsubishi?

Nissan pushed for further exploration of the second scenario, but in the end it never transpired: the engine in question was not suitable (its cast iron machining was too heavy), and the production line was not for sale. The subsequent Alliance meeting in January, 2012, validated the design of a new engine under Renault's responsibility. Finally, a completely new powertrain (engine and gearbox) was developed. The scenario for strategic convergence of Renault's and Nissan's two product plans had been stabilized.

On a technological and industrial level, the principles of a single platform and a common engine were thus agreed upon. However, beyond this in-principle consensus, questions remained on the details of the objectives to be achieved, in terms of both economic aims and industrial scenarios (which plant to select—the Alliance plant at Oragadam or the Nissan-Ashok-Leyland joint venture at Pillaipakkam?).

1.5 Convergence Driven by XBA

This exploration step highlighted the contrast between a "top-down" approach, supported by economic data and the holistic strategic thinking of the two companies, and the "bottom-up" approach promoted by Renault stakeholders involved in the project. It was clear that the XBA project, whose identity took shape in a clear and credible manner within LCI, drove the platform scenario, absorbing the Datsun project as an essential addition to obtain scale effects for volumes and the sharing of required investments such that the economic equation "passed." The Nissan initial vision for global growth has been to relaunch the Datsun brand and build an extensive catalogue via a series of local adaptations from existing vehicles. The scenario of a common platform gave the Japanese side of the alliance an opportunity to enter the sub-four-meter segment in India—and, thus, bounce back after the unsuccessful attempt at self-sustained development with Defiance.

If this project carved out a niche for itself in Nissan Product Planning as a medium-term step after some short-term launches of derivative vehicles (such as GO in India and DO in Russia), it was definitely to take advantage of the opportunity provided by the Alliance rather than to renew a solid strategic commitment by the Japanese company. Nissan never actually experienced the success of the Entry range, and its engineers did not believe in, or have any real interest in, the possibility of a low-end technological investment.

In contrast, Renault's recent experience was that the major price and cost breakthrough introduced with Logan were successful in terms of both profitability and sales volumes, as surprising as it may have seemed *a priori*. The Entry program made a valuable contribution to Renault in the difficult years of 2008 and 2009, and the range already represented more than one-third of Renault-Dacia-Samsung group sales in 2011. This program gave Renault the inspiration for intercontinentalization outside Europe, thus increasing the international proportion of its turnover from 23% in 2004 to 43% in 2011.

Behind these figures lay the strong presence of the Director of the Entry range, Gérard Detourbet. He convinced top management to grant him charge of the new product: his past experience proved his extensive knowledge of the automobile business, as well as his ability to learn the realities of markets and negotiate with suppliers and dealer networks. The fact that he was available to make the project a reality was a major factor in its favor. This disparity in motivation between the two partners would reverberate throughout the project, with Nissan resisting certain innovations strongly, and Renault to a lesser (though far from negligible) extent.

The identity of the Renault XBA project, as it appeared during the "pre-contract" period of October, 2012, *de facto* restricted the key options for the platform and imposed itself on the I2 project. It was particularly crucial in the management of key selling points (unique selling points or USPs), with a clear priority on price break.

The pre-contract review meeting was held at Renault in October 2012. This was an important meeting that gave the go-ahead to initiate more extensive engineering studies. Convened by Renault Product Plan Director Philippe Klein, the meeting concluded with the words: "Despite very good momentum,

XBA Pre-Contract cannot be granted without the right economic maturity and without a clear view of development with Nissan." In addition, it laid out the conditions to pass this milestone: "Convince Nissan to invest by sharing Renault view of profitable business; define the launch sequence for XBA and I2, secure the budget, and specify clearly Nissan economic contribution."

This indicated that, more than six months after the formal agreement on the global strategy ("single platform, two products"), the Japanese partner was still not fully invested and would commit only if the project would not result in losses for it. Such a confirmation would not arrive until the next Alliance committee meeting in January, 2013, at which the XBA pre-contract was finally passed. This milestone froze the constitution of the project.

Thus, the following aspects were defined:

1. The target markets (India first, then others)
2. The product strategy and planning (see Figure 1.7 on next page)
3. The famous four Unique Selling Points (USPs):
    a. "Breakthrough TCO" (total cost of ownership, lead price, fuel consumption)
    b. "Unique SUV Design"
    c. "My connection to modernity"
    d. "Superior roominess in a compact car"
5. Technical architecture
6. Volume targets
7. Engineering outlay
8. Target profitability

The report, signed by Carlos Ghosn, concluded: "Renault pre-contract is approved knowing that Nissan approval is mandatory to achieve committed figures."

This difference in convictions on the program between the two companies persisted after the pre-contract. The subsequent key milestone, marking the final commitment of investments from both companies, was reached for the Renault XBA project in July, 2013, but did not occur for the Nissan I2 project until March 2014. This drew attention to a budget problem that arose at Nissan at that period and the lack of internal consensus on the project.

Besides the consensus on global strategic scenarios, converging in practice was one of the major and perennial work areas of the new program. It was implemented by a new unit named 2ASDU (Alliance A-Segment Development Unit). This unit was the cause of major conflict between the management of the program and the two companies. This conflict arose following the unit's confirmation by the project team as an independent decision-making structure, which reports directly to the Alliance CEO, Carlos Ghosn.

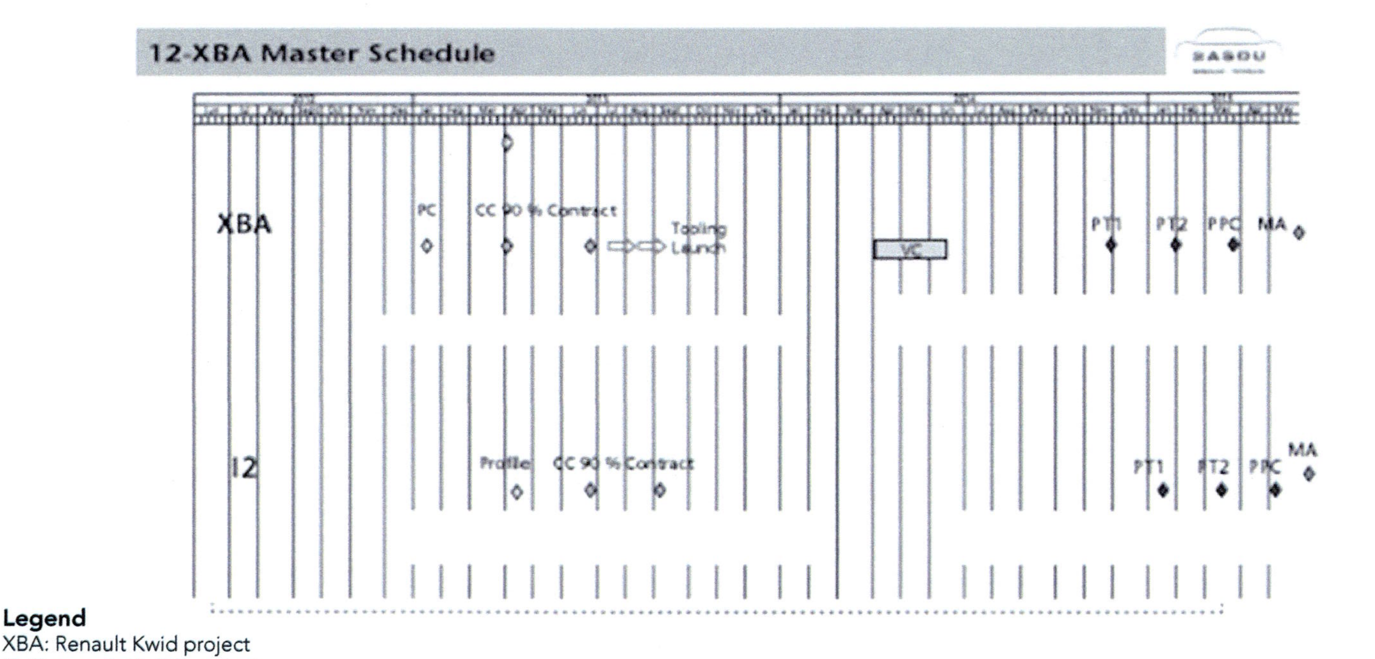

Legend
XBA: Renault Kwid project
I2: Datsun Redi-GO project
PC: Pre-contract milestone
CC: Cost contract
PT : Pre-trial
MA: Manufacturing approval

Figure 1.7 General project schedule.

Chapter Two

From Definition to Confirmation: A Unique and Independent Approach (October 2011–July 2012)

What did the new unit, 2ASDU, represent? What autonomy did it enjoy, and what action could it take *vis-á-vis* the parent companies? Until now, all common project developments had been delegated to one of the two companies, with fairly unconvincing results when it came to prompt and willing adoption by the other. Invariably, a solution developed by one partner would be diluted or revised when it came to using or implementing it within its counterpart. But that was out of the question now.

A consensus was reached on the principle of a single team, and the director identified for the platform was Gérard Detourbet, recognized by all concerned as the only one capable of taking up this significant new challenge. As a prerequisite of his involvement, he enforced a system of governance. The Alliance would govern the program, and there would be direct reporting to the Chairman and CEO, Carlos Ghosn. Nobuyuki Kawai, a Nissan engineer who had participated from the beginning to the pre-project phase, emerged as Detourbet's number two in the initial organization chart (see Figure 2.1 on next page). However, although this structure appeared in Alliance documents right from the start of the project, in practice, establishing its authority was difficult and, as we shall see, advanced only in small increments until a crisis struck in June of 2012.

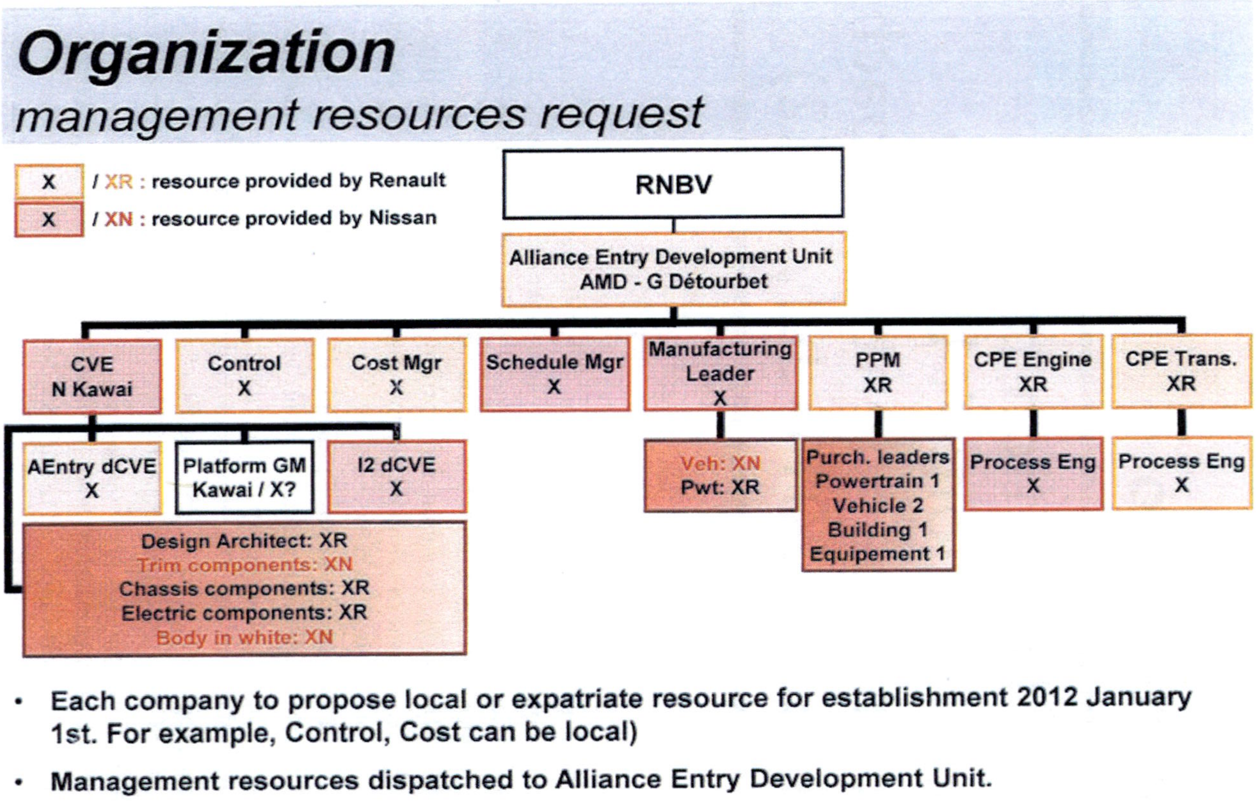

Organization
management resources request

X / XR : resource provided by Renault
X / XN : resource provided by Nissan

RNBV
Alliance Entry Development Unit
AMD - G Détourbet

CVE
N Kawai
Control
X
Cost Mgr
X
Schedule Mgr
X
Manufacturing
Leader
X
PPM
XR
CPE Engine
XR
CPE Trans.
XR

AEntry dCVE
X
Platform GM
Kawai / X?
I2 dCVE
X
Veh: XN
Pwt: XR
Purch. leaders
Powertrain 1
Vehicle 2
Building 1
Equipement 1
Process Eng
X
Process Eng
X

Design Architect: XR
Trim components: XN
Chassis components: XR
Electric components: XR
Body in white: XN

- Each company to propose local or expatriate resource for establishment 2012 January 1st. For example, Control, Cost can be local)
- Management resources dispatched to Alliance Entry Development Unit.

Figure 2.1 Target organization for the project as of November, 2011.

2.1 Formation of the 2ASDU Team

One of the key decisions taken at the Alliance Board meeting of November, 2011, was to appoint the principal members of the team. This had to be completed as early as January, 2012, in order to meet the deadlines, which were already looking very challenging.

At Renault, Gérard Detourbet appointed the core members of his team fairly quickly. His authority and engineering knowledge helped him select experienced and motivated associates for this unusual mission. Most he knew personally, and many had even worked with him on the Entry program. He considered all of them competent and qualified to find resources "at the corporate level." Others were attracted by the prospect of developing a new engine or gearbox from scratch—an exceptional challenge in a professional career, since these units' lifespans are measured in decades, and engineering is usually confined to product enhancements.

At Nissan, team recruitment was much more arduous, for several reasons. First, the project was Renault-led, giving Nissan less visible ownership. Second, Nissan's engineers had little or no interest in entry-level technologies. Finally, for the Japanese expatriates, India was not really seen as an attractive destination. As one noted, "When the Japanese leave the Indian offices, they actually celebrate, which isn't necessarily a very good sign." In the end, apart from Nobuyuki Kawai, the Chief Engineer of the Nissan vehicle project, the team was populated largely with staff from Defiance, the Indian engineering company that had previously worked on the I2 project.

The next Alliance meeting in February, 2012, affirmed the difficulty of identifying the resources required at the effective start of development. The Program Director made it known that the project was already lagging behind and would not meet its ambitious launch dates. ("Start of production" for I2 was initially scheduled for March, 2014, but was postponed to January, 2015, at the Alliance Board Meeting (ABM) held in November, 2011.) Out of 24 team positions in the organization chart, 10 were still empty as of March.

In March, the recruitment issue was finally resolved when the team was completed—predominantly with Renault rather than Nissan managers (74% and 26%, respectively). (See Figure 2.2 on next page.)

The formation of the engineering team under the managers was also sensitive. By February, 2012, there was still a shortfall of 43 managers and engineers out of the 146 envisaged, and additional expatriates from Renault and Nissan were already asking to be assigned. This staffing issue was coupled with the problem of the skills of the recruited engineers and technicians.

The 2ASDU structure created in India to develop the project was radically innovative for both Renault and Nissan. It was the first time that either of

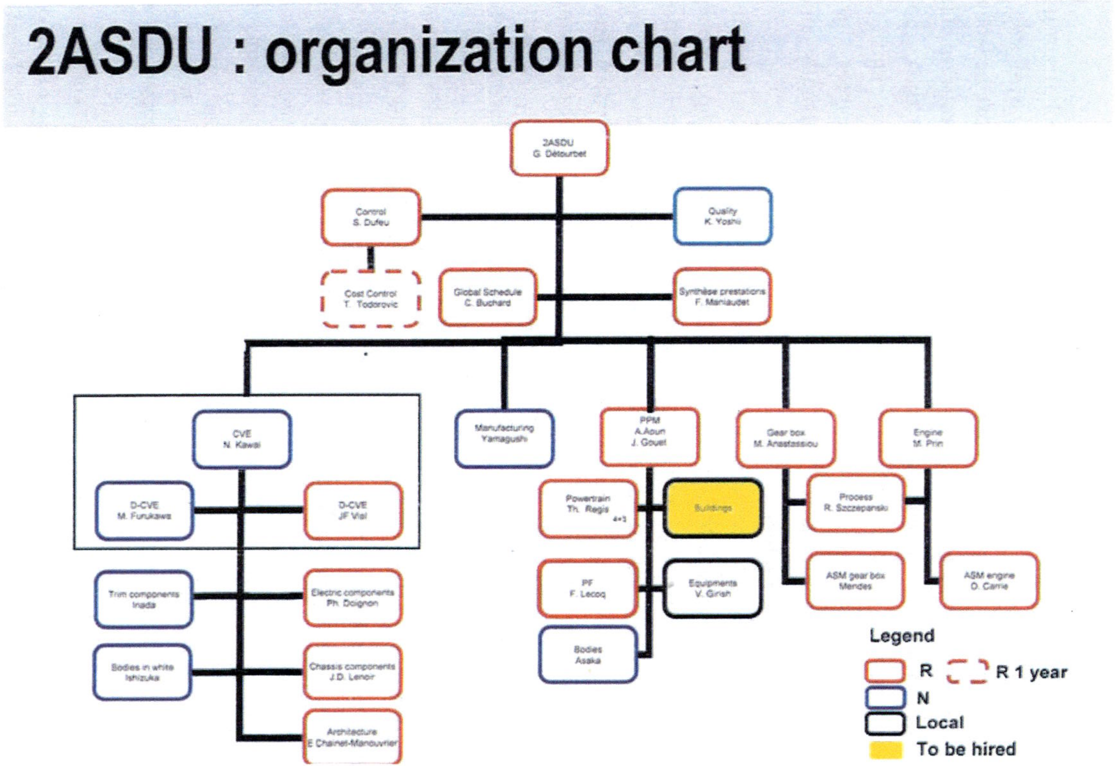

Figure 2.2 Project team in March, 2012.

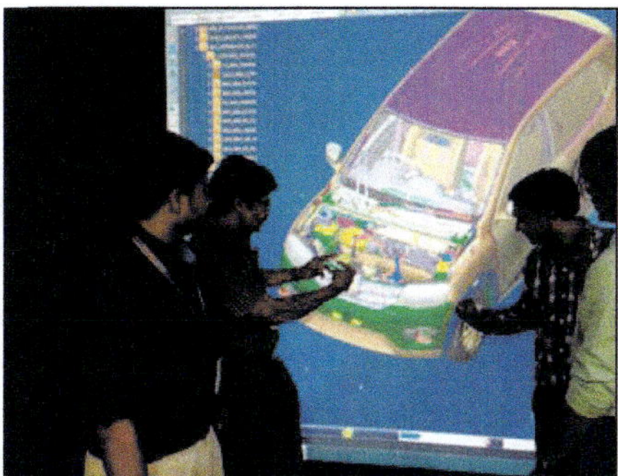

Figure 2.3 Engineering work meetings at 2ASDU war-room at Chennai. (© Renault, reprinted with permission.)

the parent companies had designed a new platform outside their home countries (see Figure 2.3). It should be remembered that even though the Logan was developed on a remote project platform, it was still close to the French Technocentre at Guyancourt (a Paris suburb). In this case, the project platform was located close to Renault-Nissan Technology and Business Centre India Ltd. (RNTBCI), the Indian Alliance unit at Chengalpattu, a suburb of Chennai. RNTBCI is a business center (information technology, cost control, engineering, etc.) with about 3,000 employees working on various projects, primarily the development of Nissan and Renault products for India. So far, these had been derivatives of Nissan or Renault products that were simply "Indianized," both commercially and industrially. RNTBCI also provided the Indian middle managers and engineers who made up the majority of the 2ASDU project workforce, along with the personnel from Defiance.

However, the development of a new platform, a new engine, and a new gearbox, which were still new to the accepted standards of the brands, required design work that was completely different from adapting an existing product locally for other markets. The local engineering skills of the Indian team at RNTBCI and Defiance were hardly suited to such a thoroughgoing redesign. "It's not just about having engineers who can design the parts. It's also important to understand why a certain part was designed in a particular manner, how it will be manufactured—and the experience of adapting models for the local market doesn't prepare the engineers for this," said Jean François Vial, 2ASDU chief vehicle engineer for the Kwid. "Fortunately, the vehicle was so simple that the expats could be involved right up to the detailing of the parts, and we could progressively train the Indian engineers in parallel." Ultimately, the 2ASDU team comprised around 20 expatriates (mainly French) with high levels of expertise not only in the engineering domain, but also in purchasing, who guided the Indian engineers and buyers.

2.2 Confirming the "2ASDU Way"

In the event, on top of the challenges of forming the team, embarking on the project in the spring of 2012 was particularly difficult. The new expatriate team members had not yet settled down in Chennai, and staff from both Renault and Nissan carried out mostly random and uncoordinated initiatives. In June, the Program Director made a harsh evaluation: "Not enough work done together; ways of working and methodologies (Nissan/Renault Entry) are very different and they do not converge; the project basic objectives are apparently shared but not really in the execution; I2 profile stress deviate too much Nissan team from urgent central tasks under 2ASDU responsibility; the orders come from everywhere; some global objectives are not fully shared or [are] misunderstood between Renault and Nissan, which create huge difficulties and especially Global vs. Local, cost USP as the first priority." He concluded: "Team is not yet operational (. . .) and does not share 'one-way thinking.' 2ASDU is not autonomous neither for management nor for methodologies. Up to now we have two months of delay."[1]

Based on this assessment, the Program Director called an emergency meeting with Carlos Ghosn at the end of June. The Director threatened to resign, something he had done several times in his career to resolve ambiguous situations. The Chairman set up a meeting during that week with senior management from Nissan—since it was clearly the Nissan side that was dragging its feet in dealing with the "Detourbet method." During this meeting, held at Nissan's headquarters in Japan, the 2ASDU Director presented senior management with his hard-hitting evaluation on the disparity between the formal commitments of the committees and practice on the ground. The action plan put forward took the form of three "requests," which were in fact non-negotiable prerequisites that highlighted the deep-rooted disparity between superficial consensus and resistance in practice: "(1) Share the general objectives (of the program); (2) accept the 2ASDU working way; (3) modify vehicle engineering department organization."

The objective of the meeting was to elucidate the autonomy of 2ASDU in light of the situation at the time: in practice, the Nissan members in particular were reporting to their original engineering department, not to the management of the program. At the same time, Detourbet also took direct charge of all platform activities; previously, chassis reported to Nobuyuki Kawai, the Nissan engineer, whose responsibility was reduced to the upper body of the Nissan vehicle, the I2 (the future Redi-GO). To remove any ambiguity, a poster on the organization chart of the proposed team proclaimed: "No more Renault

[1] 2ASDU internal document.

or Nissan but only 2ASDU people." This statement fulfilled the demand for a project identity under Detourbet's leadership that was completely independent from the parent companies.

Apart from the internal organization of the team, Detourbet also clarified the details of the contract covering product and scheduling responsibilities at the two carmakers and in 2ASDU: the parent companies determine volume objectives, product general definition, technical main characteristics, styling, milestones, etc.—in short, everything in the pre-contract document that had been approved by the Alliance meetings. It was the responsibility of the project—and thus 2ASDU—to provide answers to these objectives—for instance, "optimize with full and entire autonomy: technical specifications, product secondary characteristics, (and) manufacturing processes."

During the crisis meeting, the Nissan leaders formally accepted the preconditions laid down by Gérard Detourbet in the presence of Carlos Ghosn. This open confrontation—rare at this level of senior management, and even more so in the "consensus" culture of Japanese companies—would play a major role in asserting the project's autonomy, the legitimacy of its operational management, and its governance. It would not, of course, prevent all conflict between the companies' respective strategies, as we shall see. However, it would expedite the conditions for resolving such conflicts, since the levels of decision making and appeal were made explicit. The crisis sent a clear signal to Chennai as well as to Japan, and even to Paris. It empowered the 2ASDU leader with the real autonomy to manage the project.

Chapter Three

Product-Process Development: Spring 2012, Autumn 2013

The debates on the governance and organization of 2ASDU were pointed because the program methodology radically broke away from the traditional approach of the parent companies. And, if Nissan seemed most resistant to the iron-fisted approach spearheaded by Gérard Detourbet, it was because it was completely in line with the one he had already implemented in the Entry program. As mentioned above, such an approach was already known to Renault, while Nissan would slowly discover its radical nature over the course of this project.

Gérard Detourbet presented the approach to Nissan staff during the crisis meeting of June, 2012:

- "Top-down" cost objectives fixed to ensure the profitability of the project based on an aggressive selling-price strategy for the target markets
- "Design-to-cost" strategy imposed on suppliers for all components, which could affect functional and technical specifications in the same way as scheduling
- Reallocation of targets in case of major obstacles, and approval from the parent companies if these challenges were major
- Efforts to continue until the target was achieved, unless approved otherwise by the Program Managing Director

- Weekly reporting on cost management within the 2ASDU team, and of course regular reporting to the Alliance board

The following sections elaborate further on these points and their impact on the project.

3.1 Target-Based Cost Control

The foundation of the approach was to start with a profitable pricing strategy. Right from the initial studies, the pricing strategy was formulated as a highly aggressive move against the two dominant players in the sub-four-meter car market (which, as seen previously, is a limitation to avoid higher taxation that has an impact on the selling price). These were Maruti Suzuki's Alto and Hyundai's Eon (see Figure 3.1).

Considering the target volumes and overall profitability of the program, this price positioning imposed objectives for maximum unit cost and entry ticket (sum of development and industrial investments).

To mobilize the project team around the principle of "design to cost," a weekly meeting was set up to compare the cost statement based on the solutions

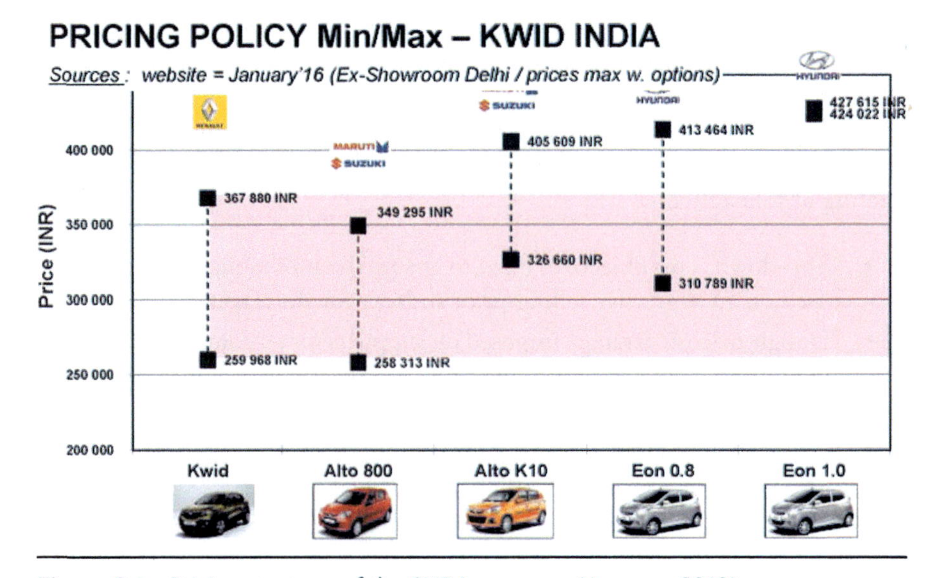

Figure 3.1 Pricing strategy of the 2XBA program (January, 2013).

in place on that particular date, and the action plans identified in order to move closer to the target. At first, the process was initiated using reference data (by extrapolating the costs of the Logan, redesigned to the parameters of the new product). It then moved on to using more accurate data for the designed parts, estimates from suppliers, and finally, contracts for purchasing parts (which constituted about 85% of the manufacturing costs).

The process of converging on the cost target took priority over all other criteria, particularly deadlines. Thus, suppliers were involved before the dates set for bids, so they could propose their ideas for saving costs before functional and technical freezes could impose irrevocable constraints. Correspondingly, contracts were only signed with suppliers once the target was achieved.

No limit was set before examining the action variables for converging on the target cost. Changes to the functional specifications might be justified by referring to the local competitive reality of India, a technical specification, a new product solution, a manufacturing process, a change of supplier, etc.

The regular weekly meetings, the methodical exploration of problems in detail, and relentless pressure from the Program Director had a powerful effect in terms of converging on the target. In fact, the ability to generate cost-saving plans constituted a key performance indicator of the project team members and the suppliers involved. A week passing without a new improvement was considered a failure for the manager concerned. "When we attend Gérard's meeting and we have nothing to report—or, worse we disclose that costs have risen—we know it's not going to go well," confided a team member. Any stakeholders who failed to understand what was expected of them were quickly removed from the project. Conversely, Program Management was closely involved in removing obstacles identified by team members—if they had not succeeded in doing it alone.

Initially, the disparity between the initial ("bottom-up") estimate and the target was striking: thirty percent had to be saved over the costs extrapolated for the Logan to reach the target set for the Kwid.

In the initial phase, observed costs began to increase when more accurate, comprehensive, and realistic estimates for the initial drawings and supplier feedback were used, in place of extrapolated technical estimates (see Figure 3.2).

From this point on, "design to cost" started to have visible effects, and the cost estimates for each scope edged toward the target. The graphs below show this progress over the course of pre-contract (January, 2012), an intermediate meeting, the contract-freeze meeting (July, 2013), and the forecast at the same date, considering the economy plan targeted.

Altogether, the implementation of "design-to-cost" methodology cut production cost to almost half that of the reference vehicle—the Sandero, the

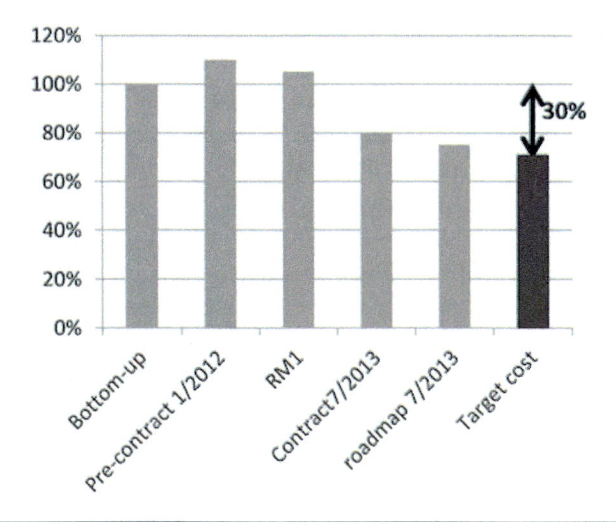

Figure 3.2 Change in costs from pre-contract (January, 2012) to contract (July, 2013).

smallest vehicle in the Entry range and the cheapest vehicle in the Renault range at that time! (See Figure 3.3.)

We saw huge gains for Purchasing as a result of these cost reductions. This was logical, given that 85% of the vehicle's value was purchased from external

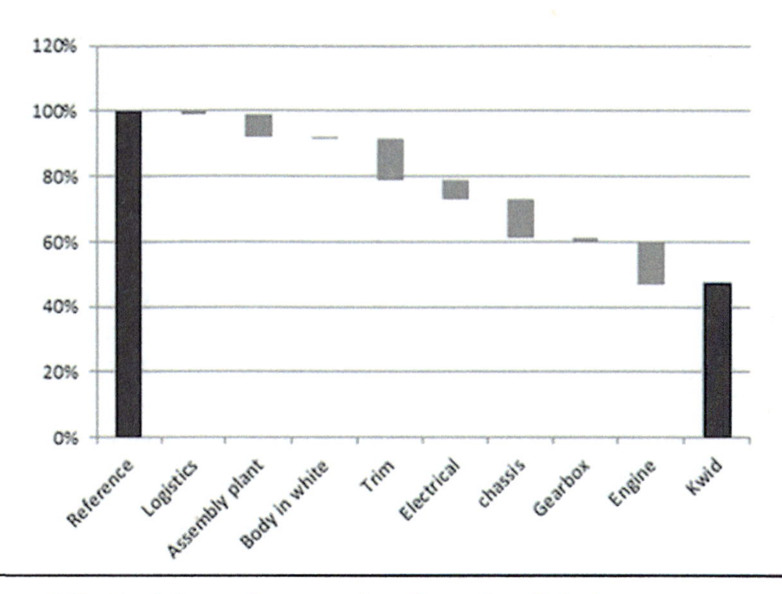

Figure 3.3 Breakdown of cost earnings for each technical area.

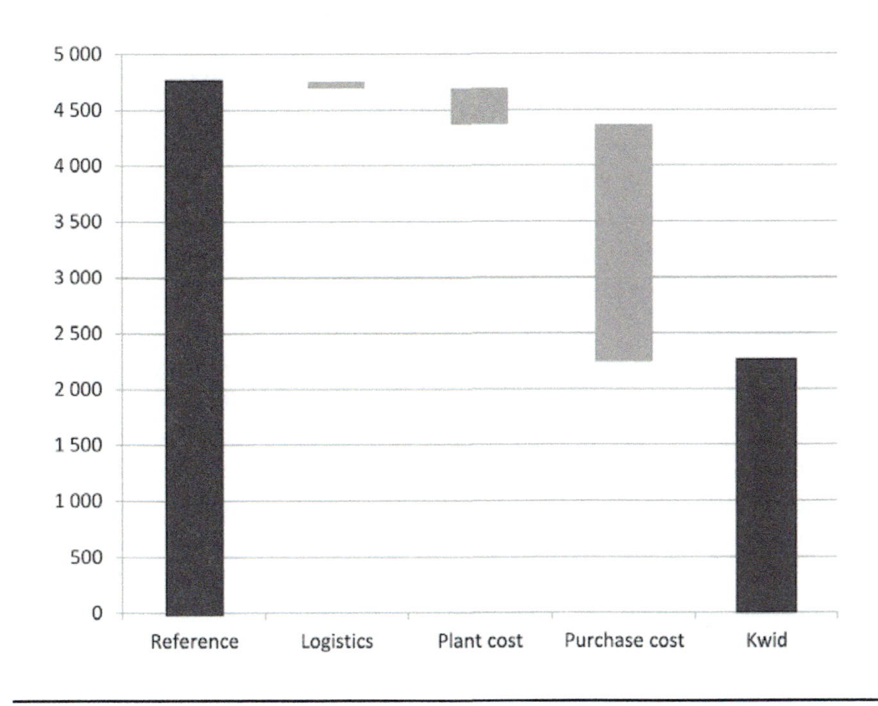

Figure 3.4 Breakdown of cost earnings in comparison to the reference vehicle.

suppliers (see Figure 3.4). By itemizing these purchasing costs for each technical area, we obtained this breakdown.

3.2 Adjusting Specifications According to the Local Market

Now let us see which action variables were mobilized to obtain these savings.The first key variable was the performance of the new product in relation to the competitor for the target market. The term *localization* generally refers to the installation of production plants; we forget that the other side of the coin is the ability to adapt the product to the conditions specific to the local market. In this case, 2ASDU had to secure a viable competitive position against the two heavyweights of the sub-four-meter segment: Maruti and Hyundai.

Once again, the project took advantage of the strategy that had proved so effective for the Entry program. We can summarize it in three principles:

1. Maintain the price (and hence the cost) as USP number one.

2. Link this price break with some other key selling arguments that differentiate from competitors (here, the exterior SUV design of the Kwid, which clearly sets it apart from the Alto and the Eon; best-in-class storage space; and an array of communications [connectivity], nonexistent at this level and conferring "most-sought-after" appeal for status-conscious customers.
3. Be just equivalent to the competitor products for all other specifications. Of course, this last condition essentially amounted to reducing the cost of the vehicle, including the user cost, by targeting improved fuel economy (km/liter in India).

The attractive style and large interior space of the vehicle, resulting from architectural and design choices, determined its basic genetics. However, the third principle had yet to be deployed in terms of the overall design of the car components. It began with a careful analysis of competitor vehicles, through systematic disassembling and reverse engineering aimed at teasing out the causal links between local technical selections and the performance achieved.

This powerful analysis raised challenges for both Renault's and Nissan's design standards, which had been successful many times but were completely out of sync with the standards being used for the target segment in India. In light of these disconnects, negotiations with the firms' respective product departments became strained. The dilemma was: should we comply with the Indian mobility conditions as experienced by most of the Indian target market? Or should we comply with the standards incorporating the identity of the brand in the European markets? Also, how do we conceptualize the versions to be derived for other emerging markets, such as Brazil?

For 2ASDU, the priority for forthcoming arbitrations was clear. However, enforcing it on those departments that protected the brand identity proved difficult (whether it was Product or Engineering that made the technical selections). One of the challenges was benchmarking performance in order to be competitive in terms of meeting the requirements of the Indian market. "We sent an Alto to Corporate so it could test and benchmark comfort-handling-braking," explains the customer requirements engineer, owner of the "voice of the customer" within 2ASDU. "The verdict was loud and clear: this vehicle was unsellable—despite the fact that it was the top-selling car in India!"

The example of door mirrors vividly illustrates these difficulties, revealing how important it was to establish sufficient autonomy in the project through a system of governance for the two parent companies.

Anyone who has ever experienced Indian traffic will likely remember it as a hair-raising experience—a cacophonous, inextricable tangle—"every man for himself" rather than the common rules of the road is the guiding principle. Even everyday driving on Indian roads involves frequent lane changes, accompanied by horn honking, demanding constant alertness and dexterity to respond to whatever is happening up ahead (and even to either side). What lies *behind*, however, is of the least concern to anyone. Door mirrors are seldom used, if at all, and are often simply folded in to the body to gain precious inches for squeezing through impenetrable traffic. Often, they are simply absent—casualties of previous, ill-advised squeezing attempts. In this context of use, the application of the project principles gave a clear answer: door mirrors compatible with the local regulations and the style of the car, certainly—but nothing more. A strict application of the "bare minimum" theory was to be applied to everything that was outside the four unique selling points (USPs).

However, this minimalistic option was rejected by Renault's Corporate Product Department. For Renault, adjusting door mirrors from inside the vehicle represented a functionality that was inherent to its brand identity. Downgrading it was tantamount to downgrading the image of the brand, which was unacceptable. The conflict began brewing from the moment the decision made by Corporate was known. Following some initial bittersweet email exchanges, envoys from Corporate Product came down to Chennai to try to persuade the Program Director. On arrival, they received the sharp advice to come more often so they could experience the local traffic. The topic was considered a friction point, as stated by the Product Director in Renault reviews of the XBA project. It even appeared on the agenda of an Alliance review overseen by Carlos Ghosn. The response was swift and decisive: the Chairman was surprised such a point had reached this level of decision making, since it so obviously lay within the Program Director's domain of responsibility. The message was clear and set a precedent, showing, *a posteriori*, the importance of clarifying the set governance during the institutional crisis sparked by Gérard Detourbet in Spring 2012.

3.3 To Save a Few Rupees More

Beyond adopting "bare minimum" functionalities, the challenge of cost could only be met by painstakingly redesigning every component of the car. By and large, incremental "Kaizen" innovation and radical innovation are often opposed. The approach used for the Kwid, as had been the case with the Logan, was something different from both.

On one hand, the radical innovation approach generally focused on a limited scope in which spectacular changes were implemented (for example, electric engines vs. internal combustion engines), whereas mainstream solutions were reused elsewhere. The Kwid approach involved no significant disruptive technologies, but all the components of the project were entirely called into question, and the challenges were extended in a comprehensive manner to the entire scope of the project: product, production process, supplier selection, logistics, and sales and marketing.

On the other hand, the Kaizen strategy demands incremental development, a "continuous improvement" of the existing situation. In the Kwid case, it involved a systematic and radical questioning of the industry standards in place in order to adapt them to the specific objectives of the project—reducing costs and achieving performance relevant to the Indian market specifically. This extended from assembly, sales and marketing, etc., right down to those parts that were, *a priori,* the least significant, or those that had long been considered fully standardized "commodities." "We pushed value engineering to its limits," recalls Jean François Vial. "Car wheels have four nuts; when we calculated using the dimensions of an XBA wheel, we could only fit three. Only, with three nuts, you have a problem during assembly, since it is more difficult to place them. So we had to design a tool to enable assembly."

We suggest the term *fractal innovation* to describe this approach. It was uniformly distributed throughout the project and operated at every level: from the overall sizing of the project to determining the features of each element; from the diameter of the cables to the characteristics of the screw-driving machine in the assembly line.

The deployment capability of such an approach went hand-in-hand with the Program Director's management style, but was further enhanced by the core members of the team. This management style involves a number of demanding conditions.

First, it requires a deep technical understanding of the whole product/process scope, without which taking on these challenges could be risky.

Second, it needs a detailed understanding of the strategic issues of the project, which allowed the team to challenge the standard levels of service required by Renault's corporate specifications. Such standards reflect learning about Western markets and are often incompatible with the requirements of the target local markets.[1] "Many of the design rules are based on experience," says Jean François Vial. "For instance, the height of the seat track. Why do we use this standard? Can we change it to save costs? We examined the principle behind its definition, which everyone had obviously forgotten. In the end, we discovered that

[1] These incompatibilities with market norms may often be small, but they can also be very numerous. The Indian market was very demanding in terms of certain requirements, such as the horn or the clutch (due to Indian driving habits), and also in terms of sustaining the commercial livelihood of the range. It was essential to deliver frequent innovations that could sustain the "product conversation" with consumers and keep the spark of interest alive. Without such innovations, sales soon fizzled out—as was the case with the Lodgy.

originally, the constraint of the specifications was that 'it should not be possible to insert a Bic lighter inside the track'!"

Traditional decision making on the project was both hierarchical (the Program Director was directly involved in the assembly line when he or she saw a problem during daily visits) and functional (recommendations from expert functional departments were regularly challenged). But this was broken down by a much more hands-on management approach, which we will call *intrusive management*—manifestly disruptive for decision-making processes set up in large groups. This is particularly true of Japanese firms, in which respecting hierarchical authority associated with the tedious *nemawashi*[2] process constitutes the basis of decision-making.

Intrusive management depends on the authority to work on the root causes of problems at both the grassroots and the organizational level. For this project, this authority was established on the basis of the professional identity of the key team members, their commitment to the aim of the project, and the organization of governance. At Renault, the importance of past experience with Entry played a pivotal role. But Nissan was a different story. "I was struck by how far Nissan engineers resisted our suggestions," recalls Gérard Detourbet. "They can refuse for a year without budging, with no explanation, regardless of your arguments. One example amongst many is one of the manufacturing scenarios we studied. In this scenario, we had an assembly line dedicated to XBA. The roller-type test bench at the end of the assembly line is a standard in Nissan's assembly process. But it is expensive, and could not be useful for our product. I understand the issue; I've managed an assembly department."

"The roller-type test bench is used to test ABS and ESP, but they didn't exist in our car. So I asked for the equipment to be removed. I went back two months later and saw that it was still in the drawings. I escalated it to three levels of hierarchy responsible for assembly lines in Nissan Manufacturing, but they didn't know why there was a roller-type test bench in the standard. I told them, 'I'm the one who is paying for the investment, and I don't want this bench.' There was no response. I went to see the big chief of the manufacturing process in Nissan, the Lord in person. I told him I could not get a process to match the simplicity of the product. I have pitches of six meters while I have a car less than four meters in length, and a roller type test bench when I didn't need one. To which he replied, 'I have an answer for you: *I* know why we need

---

[2] In Japanese means an informal process of quietly laying the foundation for some proposed change or project, by talking to the people concerned, gathering support and feedback, and so forth . . . Its original meaning was literal: digging around the roots of a tree, to prepare it for a transplant. Accessed January 21, 2017, at https://en.wikipedia.org/wiki/Nemawashi.

a test bench.' Fortunately, in the end we didn't make a dedicated line for the Kwid. Otherwise, I'm sure we would have had a roller-type test bench!"

The design work for the electrical/electronic field of the vehicle is a good example of managing fractal innovation. "On the whole, it went well," says Philippe Doignon, the Scope Manager who was present right from the start of the project. "Overall, we reduced the cost of cabling by half, from 6,000 to 3,000 rupees [€80 to €40]. We also minimized the weight. The work involved was very creative. The group studied 250 to 300 ideas—some at two, five, or ten rupees [2.5, 6.5, and 13 Euro cents, respectively], and some even at 100 rupees." Although the scope seemed simple, it was *a priori* highly influenced by external constraints. The group engaged three suppliers: an Indian supplier who was already supplying a competitor he was fully familiar with, a European supplier who had already worked on the Logan and understood the "design-to-cost" approach, and a Japanese supplier.

The group started off by analyzing the competition: Suzuki's Alto, which was the benchmark. It soon emerged that the Alto did not have the same specifications. Both Renault and Nissan rules require fastenings to be fixed every 10–15 cm. Why were these rules in place when the constraints for Indian competitors only required a fastening every 20 cm, which could save two or three fastenings? "Corporate had no answer when we came back to them on the reason for this technical specification," explains Philippe Doignon. "The company had forgotten the reason behind the standard. So they agreed to waive it if we showed that it could work." Such admissions are not always easy to make. "For connectors, we looked at what was in the market. In India, none of the horn switches in the market were waterproof—but this was prohibited in Renault. Approving a new connector would take two years. We went for an off-the-shelf solution from a Japanese supplier based in India. We had to convince [Corporate] that getting thousands more cars on the Indian roads was worth two years on the test bench at the Lardy test center in France."

This example also demonstrated the importance of a microanalysis of each part, no matter how small, with an understanding of all the systems in which it would be used. This approach revealed the relatively few levers available for saving costs on the alternator, which would otherwise have put a strain on the whole electrical system for the sake of reducing electrical consumption. The team set a limit on the maximum power of the fan, which was not to be exceeded. It modified the dimensions of the radiator grille by 17 mm so that more air could be sucked in at the same power. It also changed the shape of the blades for better efficiency at the same power and practically the same cost. "We saw that in Japan, Honda uses less powerful and smaller engines in its higher-end cars, which helped us leverage our options," said Doignon. On cable ducts, the group found that if the section size was reduced, the cables inside overheated. Thus, they had to stay within acceptable limits—particularly for sweltering India. "We had to modify the circuits, so that the key areas weren't overexposed to heat. We don't normally do this, because it is a long and cumbersome process. However, we saved 2 kg of copper and the cabling was half as heavy as in a Sandero."

This example testifies to the breadth and depth of expertise required to implement fractal innovation. "Wire harness is a domain that seems less technical, but in fact requires an understanding of the whole vehicle—everything that consumes power, and everything that generates electricity. We got results via a snowball effect: once we improved the performance of the fan, we could shrink its motor. Having minimized power consumption, we could reduce the diameter of the wire, alternator size, etc. Ultimately, we could use an alternator that weighed 3.4 kg–1.5 kg less than the one we initially targeted."

Alongside these fractal innovations was a more technically significant innovation that delivered major cost and weight savings. This was the integration of two electronic control units: the engine ECU, and the one that managed the automated manual transmission (basically a robotized gearbox). This innovation, designed in close cooperation with the supplier who provided both units, saved €30 on the powertrain as well as reducing its weight.

All these examples show the close relationship between saving costs and minimizing weight. By 2013, the car weighed 640 kg overall, making it far lighter than either the Alto or the Eon. As Gérard Detourbet put it, "We had never engineered a project by reducing the weight *per se,* but by working on saving material, finding the best solution to perform the function, we created a virtuous circle: cost savings helped us shed weight, which led to better performance in areas such as traction, handling, and fuel economy." Unfortunately, this was not the case for acoustics and impact resistance, as we shall see.

The organizational conditions for deploying fractal innovation were also clear: mobilizing leading-edge skills; systematic analysis of existing solutions throughout the automotive world; ability to mobilize networks that could access skills not available inside the group; and justifying solutions when they infringed on established norms. "I was constantly in contact with Corporate, to confirm or disconfirm our hypotheses," recollects the manager for electricals/electronics. "Initially, Corporate management was reluctant for working for the project. Relationships were formed through mutual interpersonal knowledge. Then we formulated a budget. We spent an hour per month, but it was the personal network that really made it work."

3.4 Hitting Deadlines Under "Design to Cost"

How was the "design-to-cost" approach joined up with the other immovable constraint—managing the project and adhering to deadlines? How far could the logic of "saving rupees" be taken in the face of the need to hit a launch date that had been scheduled right from the start?

Optimization efforts in product development generally come up against two obstacles: rigidly sequenced activities and sheer time pressure.

First, the sequential configuration of different expertise makes it difficult to find the right balance between the many variables of the project, some of which are already frozen when downstream activities begin. Typically, the design is frozen before engineering gets involved, and purchases are only made once the relevant components' technical specifications are set; and so on.

Faced with this constraint, a new form of cross-functional coordination known as *concurrent engineering* emerged during the 1990s.[3,4,5,6] This breaks up the functional sequence by mobilizing the various departments at every stage to identify foreseeable problems downstream (typically, manufacturing feasibility issues); improve areas of optimization (especially by involving suppliers in joint-development approaches); and ensure a smooth development process through close, continuous communication between stakeholders from the design stage onwards.

However, while concurrent engineering has the virtue of efficiency, it also represents a profound break with traditional modes of organization, in which the organization is primarily structured around hierarchical relationships within various functions. Such a change was enforced in the early 1990s by heavyweight project managers who arrived on the scene as the result of bold decisions about the colocation of stakeholders from various functional activities working on the same development. However, as we have seen, over 20 years after the theoretical affirmation of concurrent engineering, the 2ASDU project was still struggling to assert its autonomy over the supremacy of functional departments—a sign of the impressive resilience of the vertical-partitioning logic in large groups.

The second obstacle is simply the limited time available to explore alternatives to traditional solutions. Given the urgent need to converge on a solution

---

[3] Clark, Kim B., and Takahiro Fujimoto. *Product Development Performance: Strategy, Organization, and Management in the World Auto Industry*. Boston, MA: Harvard Business School Press, 1991.

[4] Midler, Christophe. *L'auto qui n'existait pas. Management des projets et transformation de l'entreprise*. InterEditions, 1993; Dunod, 2012.

[5] Midler, Christophe, and Christian Navarre. "Project Management in the Automotive Industry." In *The Wiley Guide to Managing Projects*. Jeffrey K. Pinto and Peter W. G. Morris (eds.). Hoboken, NJ: John Wiley & Sons, 2004. pp. 1368–1388.

[6] Midler, Christophe, Gilles Garel, and Alex Kesseler. "Le co-développement, définitions, enjeux et problèmes. Le cas de l'industrie automobile." *Éducation Permanente*, no. 131, 1997. pp. 95–100.

within the deadline, pragmatism and the legitimate concern of quality risk will ultimately lead to a standard solution if no alternatives have been studied in sufficient depth. In product development, the inertia of the exploration process is synonymous with conformism. It is easy to imagine how hierarchical reporting processes can be agents of inertia: any choice that needs to be "validated" must "go back" through the central services of different functional departments; it must be explained to executives who are remote from the project; it must be debated in forums in which the specific challenges of the project must be forcefully championed over received wisdom.

Contrast that with the establishment of a weekly steering meeting dedicated to converging on the cost target. The choice of compact decision-making processes within 2ASDU dramatically reduced developmental inertia, thus allowing much more diverse and wide-ranging exploration without incurring scheduling or quality risks.

Later, we shall see how project-management methods must not only facilitate such fractal innovation, but also react to obstacles arising along the way and find a way around them more rapidly than standard processes would allow.

3.4.1 Converging on a Style: An Early but Adjustable Freeze

A style freeze provides the blueprint of a milestone that marks a boundary: a radical tipping point between the world of design and the world of product and process engineering, and even the world of suppliers. Before this point, engineers are not involved in issues of detailed feasibility. Beyond it, further downstream, designers keep a watchful eye to maintain the aesthetic identity that had been "frozen."

The XBA project adopted an approach that introduced flexibility in the styling process, permitting engineers and suppliers to work together on cost optimization very early on one side, and make the style freeze "adjustable" on the other side, to preserve a degree of freedom to continue the obsessive pursuit of cost savings while preserving high design standards.

This desire to involve engineering very early on was the key flashpoint between Gérard Detourbet and Corporate Design at the beginning of the project. In July 2011, when Carlos Ghosn endorsed the film showing the convergence mock-up in Indian traffic conditions, Detourbet resigned from his previous position as Director of the Entry Program to dedicate himself to the new project. He believed that the design phase was complete. However, Corporate Design launched a competition based on the convergence mock-up to finalize the style of the car. Detourbet was furious, because he feared that re-opening

Figure 3.5 Pankaj Dhaman, Designer (left), Serge Consenza, Style Designer for the Kwid's exterior (center), and Laurens van den Acker, Renault's Design Director (right), meeting to finalize the design at the corporate designer center for exterior style. (© Renault, reprinted with permission.)

the project would merely delay the start of optimization work. This, in turn, could jeopardize the compromise that had been struck based on a mock-up with promising economic potential.

In fact, this competition was a standard phase for the designers, providing an opportunity to refine the exterior style without compromising on cost. It led to major changes to the bodywork and wheel arches, giving the impression that the wheels were larger than they actually were. It also gave the car a cubical style, shaping it into the "Baby Duster" everyone desired (see Figures 3.5 and 3.6).

This strategy of an early but adjustable style freeze would only work if Corporate Design could subsequently find solutions that maintained or even enhanced the stylistic qualities of the car while also improving its industrial feasibility and reducing its costs. Thus, the Corporate Design function was assigned the role of a solution provider, rather than a "protector" of the frozen style. Here, the XBA project took advantage of Renault's expertise as an industrial designer, an internal skill developed in the 1900s and 2000s. In addition, it relied on Corporate Design resources based in India and the international deployment of the function in various studios established across several continents, which had become one of the strategic axes of the function since the

Figure 3.6 Pankaj Dhaman, Designer (left), Laurens van den Acker, Renault's Design Director (center), and Monneet Chitroda, Style Designer for the Kwid's interior (right), meeting to finalize the design at the corporate designer center for interior style. (© Renault, reprinted with permission.)

2000s. Finally, Corporate Design Director Laurens van den Acker set up a system of regular review meetings with the Technocentre under his authority in order to get design changes validated.

For instance, Corporate Design incorporated one of the standards imposed by the Program Director to save costs on stamping: no stamping range could consist of more than three press strokes. One of the major constraints for the styles of the skin, such as the stamped part of the structure of the body, was that it must not be seen or felt on the vehicle.

3.4.2 Extended Supplier Relationships; Key Dates Driven by Achieving Targets

The same reasoning governed the process of managing supplier relationships. Here, work on optimizing components is usually completed in six months: three months for the Request for Qualifications (RFQ) process, and three months following supplier selection. For the Kwid, determining the precise specifications took almost a year, and this was before contracts were signed with the suppliers. It closely intertwined a financial logic of negotiation between the carmaker and the suppliers and intense collaboration on component optimization. On one hand, the logic was simple: once the contract was signed, negotiating cost reductions was almost impossible, so the team had to arrive at a cost target before sign-off. On the other hand, very little negotiation pressure could be exerted on the supplier's margin: considering the volumes, driving down the price could quickly bankrupt the supplier. This would be disastrous for the project, which would have to find another supplier at short notice and end up on the back foot in negotiations. Instead, the main focus had to be on the intensity of upstream optimization work to find optimized product-process solutions.

3.5 The New Powertrain Plant: An Example of Frugal Industrial Investment

The quest to save costs would also apply to industrial investment—especially since, unlike the Logan, the Kwid project involved the development of a new engine and a new gearbox. These are produced at a new greenfield powertrain plant close to the Alliance vehicle plant in Chennai (more precisely, at Oragadam). This plant is largely the opposite of a "transplant"—that is, a replication of an existing plant elsewhere. Instead, the logic was to start from scratch and build a bare-bones design, rather than use a conventional approach. Those guidelines that were used were adjusted to achieve the margin; in many cases, new and unique solutions were found. "We must experiment, even if it means backtracking and additions, if we feel they're necessary," said Gérard Detourbet. "What we're doing here is what businesses should always do: optimize the answer according to the problem and not copy-paste from a guideline without questioning it." But such an approach required a very high level of expertise and total commitment, and was only possible due to the profile of the core project members.

The facility's designers aimed to take advantage of the uniqueness of the site to save on the investment without compromising on quality. To this end, they benchmarked neighboring Indian plants by adopting innovative solutions from the building guidelines of the two parent companies. "We met the 'real' locals," says Detourbet, "not the multinational companies located nearby, which usually clone the guidelines of their western plants. They had systematically sized the facilities according to the specifications of the project (simple, small, light, less diversified product)."

The result was impressive: the total investment in the plant was slashed by half compared with a transplant based on the industrial guidelines of either parent company. The time taken to construct the plant, from laying the first stone to completion, was just six months—half the usual schedule.

3.5.1 A Plant Without Walls or Doors

One of the most surprising decisions was to build an open plant (see Figure 3.7). It had no internal walls or doors, and even the external walls were partially open—clearly, an unorthodox departure from Renault and Nissan industrial

Figure 3.7 The new mechanical unit (outside, top; inside, bottom), featuring open ventilation to save costs. (© Renault, reprinted with permission.)

guidelines. The logic was very simple: walls and doors cost money, and heat is a major problem in Chennai, where temperatures often hover around 40°C (over 100°F) in summer, and rarely drop below 20°C (around 65°F) even in winter. A visit to the plant made the results plain: the ambience was significantly more pleasant than in the nearby traditional plant, where doors were typically wedged open with stones, demonstrating *a posteriori* the pointlessness of the associated investment. In the new plant, natural ventilation systems (slats) were designed for the side walls, and mechanical extractors were located in specific areas. This design also provided light, since skylights were deliberately limited to keep out the heat when the sun beat down.

Obviously, this decision was shocking, and it had to be justified. Some departments objected: what about the dust? To minimize it, the plant was surrounded by a strip planted with grass and trees. This was mandatory in any case, in order to conform to the "green" specifications of the plant, which were imposed by the local authorities. Secondly, the machining centers were enclosed and air-conditioned so the risk was low, "and if there is a little bit of dust, it's better to pay somebody to clean it once in a while," noted the Program Director.

A related story illustrates the importance of an authority with the hierarchical power and core competence to impose unorthodox or non-conformist solutions. "The site security guard met us one day and told us, 'You have to shut the building for security reasons'," recalls Detourbet. "My answer was that it was precisely his job to make sure nobody got in. There was no question of shutting it. He did not return."

3.5.2 Deconstructing the Guidelines

Comparing the Nissan plant with the new one reveals several striking differences: smaller building columns and reduced height beneath the roof. "It's not the labor that makes building this plant costly—it's the tons of steel and concrete," said Richard Szczepanski, who was in charge of industrial design of the new plant. The height of the columns was reduced drastically compared with the common building standards, and even this had to be justified. "The roof in Japan and Europe must be able to bear the weight of 1 m of snow, which is not the case in Chennai. Unless the project team is involved, the businesses apply standards without understanding why they have been put in place." This also helped reduce the cost of power supply: "Since there are no walls, all the power is supplied by a transformer 6 meters away. By minimizing the height from the usual 'cathedrals,' we reduced the length of the pipes and cables, and hence their cost."

3.5.3 Targeted Innovations

A unique solution was adopted for the roofing, which was constructed by a Korean supplier. Cladding with profiles from a single 76-m sheet were crimped in place. Thus, there were no holes or sources of leakage. It was a single-sloped roof, which also minimized the risk of leakage during the monsoon season. In late 2015, when heavy rainfall led to major floods in South India, the building came through with no damage. Another innovation was the use of a steel-fiber-reinforced concrete mix. This avoided the cost and time spent on reinforcement, and gave results without micro-fissures—difficult to achieve in a climate where cement dries very quickly.

3.5.4 Machine Installation: Compactness, Flexibility, and Manual Transfers

The floor was uniform, with no predefined positions, in order to facilitate the transportation of machines if need be.

The installation worked on the principle of maximum compactness to minimize transportation, which was always manual. Automated transportation was unnecessary for small and relatively light parts in a country where labor is so affordable. Automated control is also costly and complicated, and can cause breakdowns. The simplicity of the program in its early stages meant manual operation made sense—but automation could be added later, when the powertrain plant eventually began exporting a major part of its production during the worldwide deployment of the program.

3.5.5 The Vehicle Plant: Difficult Coexistence in the Nissan Site

In terms of vehicle production, the project was housed in the Alliance plant managed by Nissan, which had been operational since 2010 for the Micra and also accommodated the production of the Duster and the Lodgy. The challenge was to set up in the facilities without compromising the principles of frugality. Since the plant had not been designed for the program, this was a challenge, and changes were required.

- In *stamping,* the facilities were greatly oversized compared to the actual requirements of the new product. For example, the unit was fitted with German production bridges, sized for 50 metric tonnes (around 55 tons). "There are few Indian industrial companies who produce bridges of the same quality, much cheaper than those that are partially produced in Germany or by the German companies localized in India, and who charge for the brand image of the manufacturer. Besides, 30 tons is more than enough to lift the weight for the project." Similarly, the press lines could press four or five times consecutively, whereas the product was designed to make the part with a maximum of three press strokes. Thus, the challenge of saving costs on the process for the project was reduced to organizing storage for the tooling. In order to occupy minimum space, the scenario was to pile three of them vertically: one range for each surface. This solution took advantage of the decisions taken during product design, but was not in accordance with Nissan's stamping philosophy.

- For *bodyshop,* the project used a highly specialized manual bodyshop located where finished stock was currently kept. The highly automated preexisting bodyshop would have imposed a significant additional cost—not only for investment, but also for operation, due to the maintenance costs of automated lines.
- In *assembly,* the project team initially studied the scenario of a dedicated assembly line. However, this was rejected because the plant was greatly oversized at the time: the existing Nissan and Renault products had not been as well received as expected in India, leading to manifest production overcapacities. Moreover, the future left-hand-drive Micra for Europe (then exported from India) was to be produced in Flins from 2016. Thus, the project opted to use one of the existing assembly lines, adapting it to make it more efficient. The other products made at the plant were more complex and diversified, which necessitated assembly strategies such as *picking* (preparation by collecting parts to be assembled for each model based on the exact specifications of the variants, options, etc.). But such an approach might be less useful for the much simpler new vehicle and could lead to additional costs.

One of the challenges in the three workshops was to not compromise the performance of the new product's assembly line because of other processes. This required management systems to isolate the various branches, which is definitely not generally the case today. In fact, prevalent manufacturing cultures are defined by the *absence* of a cost-effective management approach—a classic feature of Japanese production units, which are "physically" managed. However, while this works well in the Japanese context, where operators optimize processes and are meticulous in their implementation, the Indian context is very different. Here, emergent discrepancies will become permanent if workshop managers do not take decisive action.

Although the project team was in control of the new powertrain plant, it still had to deal with the management of the assembly plant, for both vehicles. In Nissan's culture, an even more uncomfortable situation in terms of innovative strategies was that the Project Manager's role usually ended at the perimeter of the engineering process, since it was merged with manufacturing. This sequential, compartmentalized development process stood in sharp contrast to the 2ASDU project organization, with its tight integration and short loop for product and process components.

It was easy to understand why Program Management's position in the Alliance had been a fundamental prerequisite for imposing manufacturing options specific to the project. "When the new Alliance organization chart was published, Nissan's people could see that I reported directly to Carlos Ghosn," recalls the Program Director. "They hadn't known that before, even though it had been that way right from the beginning. From then on, I had a feeling that they listened to me differently. But there's no hiding the fact that it's difficult."

3.6 Behind the Scenes in India

While the project knew how to capitalize on the advantages of Indian localization, it also had to deal with its constraints, which were equally numerous. With work caught between the ideal of a streamlined modern automobile process and the context of "Incredible India," conflicts were significant and blew up daily.

3.6.1 No Plant Is an Island

The industrial revolution of the 1980s unleashed Toyota-ism on the automotive industry worldwide. This rationalization theory weeds out risks and defects based on the principle of production flow. The metaphor often used to describe this approach is that of lowering the water level of a lake so the rocks are exposed: tightening the flow makes the problems visible, which then raises the obligation of dealing with them in order to eliminate them. This approach fundamentally contradicts the previous perspective in the car industry, which focused on ensuring industrial performance by preserving the production system through "buffers"—stocks or safety margins that made it immune to risks.

The advent of the "Toyota Way" in the West significantly transformed not only automotive plants, but also the broader environment on which they were based. Indeed, production optimization would quickly reach its limit if confined to the boundaries of the plants: the production flow of sourcing depended on the capability to ensure efficient logistics; elimination of defects in the chain rested upon the efficiency of learning processes, which required a stable workforce and consistent schedules; the reliability of technical means depended on the quality of energy production; etc.

India has a strong industrial tradition, on which the project could capitalize. However, the broader context within which it unfolded was still that of a developing country. Trying to incorporate "Lean"[7] approaches required tenacity and pragmatism; the idea that an environment could be transformed at the wave of a magic wand was wishful thinking. Essential emergency stocks in a flawed infrastructure made the sourcing system slow and unpredictable; policies were directed at stabilizing and training the local workforce; and actions were initiated for energy operators to improve the quality of energy supply.

3.6.2 Lean Production Versus Indian Administration

One of the dark sides of "Incredible India" for an industrial project is bureaucracy and myriad, labyrinthine taxation rules. If these administrative questions were not resolved, the associated costs could erode the competitive advantage of being in the country—which, it is easy to forget, is the size of a continent. "A few components went back and forth between two suppliers before being assembled in the final assembly plant," recalls Céline Buchard, Head of Logistics. "The raw part was dispatched by the first, assembled at the second, and returned to the first, who finished it. Since the two suppliers were located in two different States, we paid the tax for every transfer. Despite receiving notifications over and over again, India still could not establish a logical approach to the VAT, so we were asked to pay a tax of 6% on the overall cost three times over! To avoid this madness, we had to set up a complicated system and negotiate it with the authorities of the States."

Such recondite complexity was a godsend for a prolific bureaucracy that seized on the slightest ambiguity as grist to the mill. As a continent, India would make an interesting case to be analyzed through the lens of Michel Crozier's theory on Bureaucratic Phenomenon.[8] At every stage, obtaining the necessary authorization to move forward was subject to the discretion and goodwill of those who "controlled the area of uncertainty" by bestowing the lifesaving magic stamp.

[7] Womack, James T., Daniel T. Jones, and Daniel Roos. *The Machine that Changed the World.* Cambridge, MA, USA: MIT Press.

[8] Crozier, Michael. *The Bureaucratic Phenomenon.* Chicago University Press, 1964.

The phase of building prototypes was particularly complex from a logistical point of view. In fact, even though 95% of the parts purchased were located in India, there was no means of designing prototypes in Chennai to build the body or carry out tests. The option selected was to build bodies in white and test them at Nissan in Japan, which was well equipped for this purpose. They would then be brought back and prototype vehicles assembled in Chennai using parts from suppliers. Finally, they would be tested in France. In such a scenario, efficient logistics was a key factor to the success.

However, the Indian scenario put major stumbling blocks in the way of this approach. On one hand, transportation was slow and irregular due to nightmarish Indian traffic. On the other, the transit routes of the merchandise fell victim to abundant and discretionary bureaucracy, which put the entire approach in doubt. "I completely underestimated the administrative complexity necessary to transport the parts," recounts Céline Buchard. "We wanted to build prototypes in the plant, but the logistics department there knew nothing about running a prototype workshop. They handled a packet of screws as if it were a container filled with engines! I was told that, to import one part, we had to go through a complicated financial transaction because of the taxes, and if the procedure was not done correctly, we could go to prison. In India, when you need an authorization, you are always told that it is very complicated. First, you need the signature of an official who does not delegate. So there's a wait. Initially, when there was some urgency, I went there alone; that was a mistake. In India, people's status is judged by the rank of those who are made to wait [to see them], and for how long. Thus, they were delighted to make me wait until late into the night, just to show others their importance. Now I know better, so I send people from the lower rungs of the organization chart. But it's still very costly: I have five people who do nothing but wait in line so that the applications are pushed forward."

Chapter Four

Supplying from India

Purchase of automotive components to external suppliers represents 85% of the Kwid's value. To meet the cost-price target, suppliers were deeply involved in the "design-to-cost" process to obtain competitive prices for components to be assembled on the powertrain unit and the final assembly line.

If Japan was the country of lean production at the end of the 20th century, India is now the most frugal manufacturing country, based on low labor costs and a strong ability to target what is "strictly necessary." Not only was the car designed in India, it is assembled there and its components are produced there, with a record level of localization from local procurement. The level of integration has reached 97.5% for the basic versions of the Kwid (falling to 92.2% for the finishing known as RXT/E4, including the most modern electronic equipment, which is imported). Admittedly, this is based on face value—that is, the value of purchases from direct suppliers to the Alliance plant. Since these suppliers buy components that may be imported, the actual value of the local contents is slightly lower (around 90%). However the extent of "Indianization" in purchasing is still remarkable—and unpredicted.

4.1 Indianization

"There must be clarity on what we call the 'local supplier'," explains the Program Director. "There are Indians, and then there are the 'fake Indians.' In India, there is a strong local skill base in the mechanical engineering field. But often,

Indian manufacturers only do business with Indian customers, while other Indian firms who specialize in doing business with Western companies outsource their work to the former. Obviously, with this switchover, cost performance is adversely affected without the arrangement necessarily adding any value. I took the time to gain access to the 'real Indians' and capture their interest, because they are not too keen to conduct business with Westerners, let alone export. Another type of manufacturer is a subsidiary of an international group, who would offer rationales and solutions found in their home country. They can be interesting when it comes to highly technological spheres, such as we had on the project."

Large multinational component makers did feature among the Kwid suppliers, but they manufactured only one or two highly targeted components. These included Saint-Gobain and Asahi for glazing; Valeo for the starter motor and EV purge valve; Bosch for the injection system and braking booster; Visteon for the air conditioning (HVAC) and radiator; Delphi for the fuel pump; etc. However, these suppliers were also subjected to the "2ASDU way" with tougher negotiation, and the success of the project lay in more than half the suppliers being "real Indians" (see Figure 4.1).

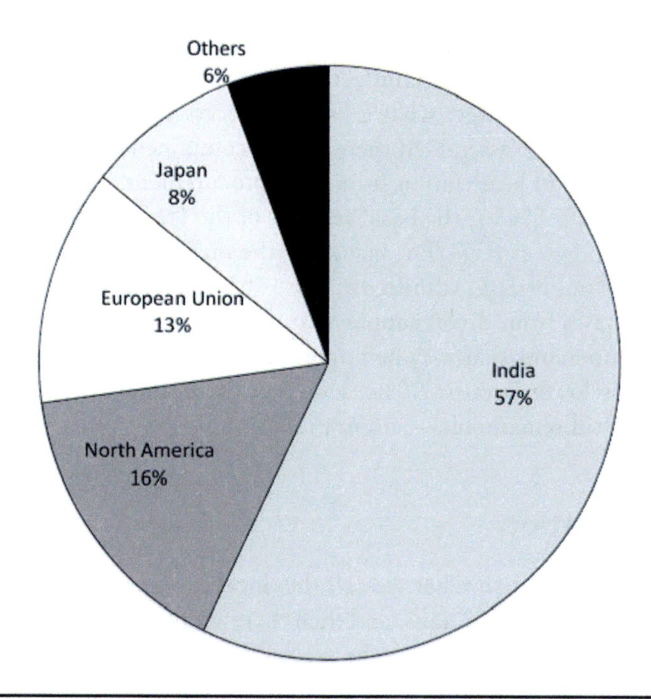

Figure **4.1** The origin of capital of the 70 main Kwid suppliers, which were all localized in India (*Source*: 2ASDU.)

This success was the culmination of a mutual learning process between Renault and the Indian suppliers that took several years. Renault developed its knowledge of the Indian automotive component industry by contacting it directly during the deployment of the Entry program in India with the production of the Logan (sold under the name Verito) by the Mahindra & Mahindra car manufacturer from 2006 to 2010. Discovering that prices were 20–30% lower than in Romania (labor costs are five times lower in India), Renault duly involved the Indians in supplying the Entry range, especially during the launch of the Duster in Europe. But Renault found it difficult to sign contracts with the local suppliers, who already had attractive opportunities with the car manufacturers who were present in India's high-growth automotive sector at that time.

The start of the production of Renault cars by the Alliance plant, especially the successful launch of the Duster (60% "sourced" in India), opened a new chapter in terms of local understanding. As previously described, the Alliance plant's products (whether Nissan or Renault) fell short of expectations, failing to achieve the projected volumes, and suppliers became hesitant. However, the presence of Renault's car buyers in India, coupled with the prospect of significant purchase volumes (production of 200,000 vehicles per year in India, plus additional volumes for export), worked in the 2ASDU's favor. Careful review of the panel of Indian car manufacturers (Maruti and Tata, in particular) discovered technical solutions that had not yet been explored by Western car manufacturers but were widespread in India. They could also identify the most competitive manufacturers in terms of costs, with the potential to develop skills with adequate support. "We worked with the suppliers of the big Indian car manufacturers: Maruti, Suzuki, Tata," explains Jean-François Vial. "As far as the fuel system is concerned, we did a carry-over of the Tata Nano fuel pump. The door locks were extrapolated from those of Alto, and so on. In this way, we could take advantage not only of the Indian suppliers' frugal design skills, but also the benefits of mass production."

Therefore, supplying was crucial for 2ASDU, and it was closely monitored by Gérard Detourbet, who had already made Purchasing (identification of suppliers, negotiations, monitoring, etc.) one of the strengths of the Entry program. He relied on a Purchasing Director (first Jérôme Gouet, then Nuno de Morais from the end of 2015) who was experienced in the automotive industry at the international level, and who managed a purchasing team of around 50 people, supervised by three expatriates deputed by Renault.

In the end, out of the 140 suppliers for the Kwid, two-thirds were already suppliers for the Alliance plant (RNAIPL), and 2ASDU had identified 46 new cost-effective suppliers. Among these Indian suppliers, some were multinationals who already had several production units on different continents, whether Europe or North America. One such was the Motherson Group, which started out in wiring in 1975 and quickly diversified into other technologies (including plastics manufacturing). It established partnerships with numerous well-known OEMs and benefited from the restructuring of the global parts manufacturing industry to acquire reputed companies such as Peguform. With a presence in 25 countries in 2014, its turnover topped six billion dollars. Motherson supplied the Kwid bumpers and several plastic parts for the interior fittings of the vehicle, including the instrument panel (dashboard). "The collaboration with 2ASDU was a chance to take a new leap, including into the design stage, because of the close relationship established with Jean-François Vial through daily meetings. This attentiveness allowed us to develop in design," explains L. Srinivas Reddy, Vice President of MATE, part of the Motherson group.

Even for Indian OEMs with international exposure, the venture was not totally risk-free, considering the financial fragility associated with growth that could sometimes be too rapid. This was reflected by the misfortunes of the AMTEK group, which was selected as a supplier of several Kwid components, and which wound up on the verge of bankruptcy and underwent significant restructuring. At other times, the restructuring of the global parts industry had repercussions for the project when the part-resale of one firm to another excluded India. This was what happened to the joint venture between Johnson Control (JCI) and the Indian firm Pricol, which had been set up to produce an innovative electronic component. The acquisition of JCI's electronic automotive business by Visteon in 2014 left its Indian partner stranded, because India was excluded from the deal. In the end, the XBA lines were transferred to Visteon.

These big Indian OEMs easily passed the certification procedures put in place by the Alliance, and several manufacturers were already part of the Alliance plant panel before the XBA/I2 project. However, this was far from being the case when 2ASDU took a chance on new suppliers, who would have been eliminated immediately under ASES (the Alliance Supplier Evaluation Standard). In such cases, continuous support was crucial in transforming manufacturers of motorcycle or moped parts into automotive component manufacturers. 2ASDU designed the parts, produced the tooling (mainly in South Korea), and bought the materials to supply the OEM. Almost daily monitoring was required to help the supplier meet the standards and fulfill its commitments.

Anash Haime was responsible for supporting suppliers and maintaining constant vigilance on the application of procedures. This true Renault representative, a French native of Pondicherry, was assigned by Corporate to the Alliance plant when it opened, before later joining the 2ASDU team. He was responsible for ensuring that suppliers complied with the Renault-Nissan procedures sanctioned by signing the PSW (Part Submission Warrant), which ensured that parts would be compliant with the Alliance standards. With a team of eight people (all Indian), this cross-departmental quality function sometimes had to think ahead to ensure that some suppliers complied with the standards. The individual monitoring of firms included fortnightly, weekly, or even daily visits for the most precarious cases. These visits allowed the team to monitor crucial developments, but each one invariably involved a reaffirmation of the same essential principles. "At each visit aimed at resolving a problem, we inspect the basics," says Haime.

4.2 Three Supplier Stories

To illustrate the approach, we present three brief examples showing the diverse trajectories followed by suppliers to the project.

4.2.1 Supplier A

Supplier A is an Indian manufacturer, located one and a half hours from New Delhi. It is a family business with 5,000 employees, international partnerships, and proven engineering skills, plus previous exposure to the global automotive industry. Supplier A had to supply a rather complex mechanical part for the new engine, which required the acquisition of new skills. "The corporate businesses of Renault had invested heavily to assist [Supplier A], with the support of RSM in South Korea, which acted as a kind of mentor," says Thomas Régis, the buyer.

Supplier A bought top-quality Italian equipment worth several million euros to supply this part by borrowing in foreign currency. However, with the fall of the Indian rupee, it found itself in financial trouble, and its employees were not paid for several months. Two months before the start of production, Supplier A was invited to meet the 2ASDU team to discuss the situation. The overcrowded meeting room contained, on one side, a smaller 2ASDU delegation (Program Director, mechanical part buyer, Deputy Director of Engine, Indian buyer), plus an Alliance plant buyer; on the other, about 15 people from Supplier A, including the CEO. The objective was to verify whether Supplier A would be able to handle the increase in mass production and specify the timing of its planned installation in Chennai. If the defect rate on pre-production was considered to be reasonable at this stage, the means to achieve the final quality objective would need to be specified.

The meeting began with a very general PowerPoint presentation from Supplier A. This "disingenuous" speech quickly irritated Gérard Detourbet, who interrupted it to firmly state the issues he wanted to deal with. First, he wished to discuss each defect noted on a batch of parts, identify the causes, and clarify the solutions being considered for dealing with it. Second, he wanted to talk about better planning for machinery installation in Chennai, where Supplier A needed to be settled in the coming weeks, against a background of rapid growth in volume and diversification in production.

The discussion turned toward the defects identified. Supplier A indicated they were still working on the problem. The CEO conferred with his engineers in Hindi before replying in English, betraying the lack of preparation for this crucial meeting. After an interruption for a visit to the assembly line, the supplier made some requests related to an agreement on the final price in mass production, problems with tooling and molds, and finally logistic issues. Detourbet pointed out that these issues had already been finalized, while leaving the door open for future negotiations.

A further interruption provided the opportunity to meet spark-plug maker Federal Mogul, a global supplier of an engine component, in the next room. This component had been redesigned for XBA/I2 to cut costs in an effort to save every last rupee. The standard version was oversized for a 0.8- or 1-liter engine, and attention was focused on reducing it. Federal Mogul could offer it to other Indian manufacturers using engines of this size to serve the market segment that the Kwid was targeting. However, the new product had not been validated, and could not be validated before the start of production of the car on a large scale. Detourbet suggested that these components should be imported from France at the early stages, and that Federal Mogul should bear the additional costs. To reduce the cost burden, he offered to make an exception to Renault's rule of not changing a component for the six months following the launch of a new model. Instead, he said he could introduce the new component after only eight weeks, or earlier if it was validated quickly. The matter was settled in half an hour.

Before returning to Chennai, Supplier A organized a ceremony in which the 2ASDU team planted five trees next to the plant (two were previously planted by a German manufacturer). Ten months later, the firm's plant (whose opening had initially been planned for before the start of production) was still not operational in Chennai, and the part was still being transported across India for use on an engine assembly line in the powertrain plant.

4.2.2 Supplier B

Relocating suppliers from the North of India to Chennai was difficult, and this issue gave rise to a rare instance of failure in the attempt to support skill development. It could even be called the case of default. Supplier B is a family business that produced mechanical parts for XBA/I2 cars, having developed ironworking skills that were lacking in South India. At the start of production, its machinery had been transferred from the North into an unfinished structure. Although located in an industrial zone, the facility was not connected to the power grid and ran off a generator. More than six months after the start of production, it remained an illicit operation that did not have official permission to carry out manufacturing activities. This improvisation was reflected in the employees' terrible working and living conditions: employee toilets had only just been built, and there was no running water, let alone a canteen. Some migrant workers slept in a hovel next to the plant.

The purchase price of Supplier B's part was unbeatable. But a visit to its Chennai plant revealed a lack of organization that was scarcely believable for the 21st-century automotive industry. Although new, the plant was woefully dirty. Material flows were unmanaged, the safety of the operators was far from assured, and the prevailing impression was one of improvisation. It was like an industrial environment from the imagination of Émile Zola. Around 60 employees were busy repeating the same movements under harsh and dangerous conditions. Their wages were below average, even though they worked 12-hour shifts, seven days a week. Since job opportunities were plentiful in the surrounding industrial zone, operators did not stay long, and staff turnover duly reached record levels. This applied equally to management: the plant had three managers within eight months. This disastrous régime was accompanied by a total lack of compliance with procedures, with materials or processes being changed without warning. Errors were common, and quality data was even falsified. This supplier would be replaced, but it took time and could not be achieved before mass production started. Pending this replacement, the 2ASDU team visited the plant daily to address the most glaring failures.

4.2.3 Supplier C

Such isolated failures should not detract from the overall success. In total contrast, Supplier C was a medium-sized European firm that produced small forgings for the Kwid (including some for export to Brazil). It was a supplier for major clients in Europe (both car manufacturers and OEMs), but was without an industrial presence outside its home country before its localization in India for the project. Supplier C initially had to join forces with an Indian partner, but this idea was abandoned after a few months of negotiation due to incompatibilities between the two firms.

In spite of the belated decision to construct a new industrial building, the deadlines were met. The plant was modern, huge, and well organized around material flows. Large presses from Europe were installed, but parts were handled manually, in contrast the automation of Supplier C's European plants. Small presses were installed for other finishing operations. Working conditions seemed favorable, with careful attention to the safety of the employees (earplugs, goggles, etc.). The work was carried out in two shifts (eight hours per day, six days per week), and the employees were paid above-average local rates. The manager was the only expatriate among some 80 employees. The contrast with Supplier B could not have been more striking.

4.3 Supplying in the Chennai Region

Historically, the Indian automotive industry had developed around two centers: the capital, New Delhi, where market leader Maruti Suzuki is based; and the Chakan corridor in the suburbs of Pune (quite close to Mumbai), which brought together a number of Indian and foreign car manufacturers. The component makers were principally located close to these two centers. Chennai has developed more recently, but the influx of foreign car manufacturers (Ford, Hyundai, Mercedes, BMW, and Renault-Nissan in particular) has now made it the number-one automotive center in the country, particularly for exports. However, its automotive component industry was less concentrated, with (as noted) a lack of certain skills, such as ironworking.

For the XBA/I2 program, the procurement aim was to secure suppliers who were based close to the Alliance plant. This prompted companies such as the three mentioned above to open new plants in Chennai. The program also stood out with an exceptional proportion of domestic content (98%): sixty percent of components are produced in the state of Tamil Nadu, mainly in the area around Chennai (its capital).

Such proximity brought many benefits, especially in terms of costs. Due to poor road networks, deliveries could take two to five days to come from Pune or New Delhi, with uncertainties about the arrival of the goods. This posed substantial risks related to disruptions in supply. Moreover, logistics companies had a tight grip across India: industrial transport costs for Chennai destinations were 15 times higher than if located in Tamil Nadu, and there were no alternatives to be had. Then there were the taxes for cross-border transit discussed above, which impeded interstate logistics flows. It made sense to invite suppliers of heavy and bulky parts to set up their businesses close to the Alliance plant. Sometimes there were delays in doing this, especially since there was a lack of confidence in the volumes announced.

Céline Buchard took on the job of managing this complex logistics management function within the 2ASDU—among other functions, including monitoring suppliers' sites. By spring 2015, she had a team of seven people—three in Chennai, two in Delhi, and two in Pune—meaning she could have someone on the suppliers' sites within half an hour in the case of any problem.

Buchard set up an original system to manage procurement in the form of what was known as the "milk run": trucks made the rounds to collect components and equipment from suppliers' sites or warehouses for transport to the production lines of the Alliance, the 2ASDU powertrain plant, or the assembly plant. In a remarkable system that is unique in Renault worldwide, the supplier provides the packages and invoices the transport, while the carmaker manages the logistics.

In February, 2016—more than six months after the start of production of the Kwid—we evaluated supplier relationships with the 2ASDU team. While there had been four or five failures among the initial panel of suppliers selected for the project (most of which dated from before the summer of 2015, when the first cars rolled off the assembly lines), the overall assessment was positive. In a few cases, there had been reason to switch supply from a single supplier to double sourcing. Constant vigilance was needed to keep tabs on volume and quality, but the outlook looked quite favorable in spite of a rapid increase in complexity for the months ahead as new challenges emerged.

First of all, the production ramp-up of the Kwid needed to be monitored to cope with the deluge of orders. In July 2016, the assembly line rolled out 2,500–3,500 cars per week. In the summer of 2016, mass production of the Datsun Redi-GO began, representing a significant change to the volume mix with the Renault Kwid. Initial projections foresaw a 50/50 split, with 100,000 vehicles per year for each model; at present, it is more like 165,000 Kwid and perhaps 65,000 Redi-GO. While parts were widely shared by both models (90% for the underbody, including the powertrain), the differences in bodywork design obviously reduced the commonality rate. Changes in the volumes for specific parts led to very significant gains for some suppliers, but declines for those who were not on the Kwid program.

Diversification of the production had to be managed both at the mechanical level with the launch of the 1-liter engine and automated gearbox, and at the vehicle level with new versions that will be rolled out in 2017. In the autumn of 2016, production of the Kwid in Brazil (XBB) will finally begin, and in the early days the car will be largely fitted with Indian components (45% references, but only 19% of the purchase value). Not all Indian suppliers wanted to export, and some were replaced by new suppliers. Eventually, purchases in India will have doubled and the complexity of the logistics carefully managed, since the flows to the Alliance plant in Chennai have been joined by those to the two Indian sites for the preparation of exports—namely, Renault's ILN (International Logistic Networks) in Chennai and Pune.

Chapter Five

Investment Decisions at Industrial Start-Up and Ramp-Up: July 2013–June 2016

With the design phase completed, it was time for the project to put its ideas into practice. After the freeze on major decisions—namely, tooling and supplier selection—a new phase of the project began. Its scale and nature changed, switching from a design philosophy managed by a relatively compact, highly competent, and motivated team to the routine operations and standardized processes of the production plants. This was a time when there was little left to gain, but potentially a great deal to lose, if the hypotheses behind the design decisions turned out to be faulty or if unforeseen events intervened. Overall, it was a tricky time, and one that could easily undermine ambitious projections if they weren't in line with reality.

5.1 Responding to Risks

In terms of risk management, the 2ASDU approach was similar to critical-chain theory as originally formulated by E. M. Goldratt in 1984: systematically eliminate all the margins locally accumulated by the stakeholders in charge of the batches of the project, and instead consolidate them at a global level; secure key resources (for example, key stakeholders and test facilities); set up a management system that could detect early on if the expected "bare minimum" will be competitive—and if not, drive forward corrective responses by anticipating backup solutions, if any were possible.[1]

For risk identification and detection, the success of the project was most strongly assured by the skills of the teams, extensive in-depth studies, and the continuation of approved solutions for the Indian market. Because the "why" had been comprehensively explored in the upstream phase of technical selections, stakeholders were rarely surprised by test results. Yet a number of hurdles still had to be overcome. For example, there was the technical challenge of lightening the powertrain with plastic oil housing; supplier challenges with almost all suppliers outside the panel; and product challenges in terms of alignment with the services of the existing competitors. Clearly, "fine tuning" the design to the bare minimum *a priori* could not be successful every time, and two domains proved especially problematic: vehicle impact and acoustics.

5.1.1 Development of Crash Safety Performance

In line with the "bare minimum" strategy, the structure of the XBA's body was adjusted according to the benchmark of its main competitors, Alto and Eon, for approval in the Indian market. The project-management team's decision was to validate the structure of the platform during adjustments by removing the first batch of prototypes lined up in the development plan in September 2012. Reducing the investment, this saved money and gained about eight months of time for development.

In autumn 2013, feedback on the adjustments on the overall body, which had previously been good, took a turn for the worse. Three kilos of reinforcements were added to the first prototypes of the competed XBA body, which were available in September, 2013. Everything hinged on the evaluation of these body prototypes, which would be tested on Nissan's test benches in Japan, because the requisite facilities did not exist in India. By 2014, feedback for durability and impact was positive once again.

During the auto expo in Delhi at the end of 2014, Global NCAP (which publishes the most well-known rankings for passive safety) conducted its own crash performance tests for all makes of Indian cars. The results were disastrous. This was naturally followed by intense lobbying for improvements, and Global

[1] Goldratt, Eliyahu M., and Jeff Cox. *The Goal: A Process of Ongoing Improvement,* 2nd rev. ed. Croton-on-Hudson, NY: North River Press, 1992. (First edition, 1984.)

NCAP's president spoke directly to Carlos Ghosn to share the poor results of the Datsun GO, which was among the worst rated. It was decided to discard the existing Indian market as a benchmark and reinforce the body to meet a more stringent impact standard instead. The car would face such standards in Brazil in any case, but they could also be met in India.

"This sort of decision, at this stage, is disastrous for schedules," noted Jean François Vial shortly afterwards. "Usually, such a readjustment in structure requires a design/validation cycle of at least a year. Manufacturing approval is planned for July 2015 . . ." A reinforcement kit was developed immediately, then optimized for a realignment introduced after the initial launch of the vehicle. After the second readjustment, the vehicle's weight jumped by 20 kg, and its cost increased by 1,500 rupees. After three months, the realigned prototypes were sent to France for tests; as Vial had noted, such a realignment cycle would usually take a year. "What helped us was that this platform could be upgraded for other markets." Moreover, the schedule could be shortened due to the "density" of the project team, and its competent stakeholders were duly brought to bear on the problem. Working together in a compact team, they could make decisions within a week or, in critical cases, on the same day. By contrast, change-management processes set up in large groups involve several stakeholders who are not always aware of the multiple authorization levels of other stakeholders facing multiple challenges and working in much slower decision-making cycles. In the end, the internal impact tests were satisfactory.

The impact on deliveries was minimal: the Kwid would be launched in October instead of July, 2015. The goal was to make up for this delay with a more "vertical" ramp-up in order to maintain the particularly ambitious (and unchanged) 2016 sales target. The production date of the realigned vehicle with improved crash safety performance would be April, 2016.

However, these "commando" processes would have to be harmonized with the "normal" start-up approaches of the plant—approaches that worked on the assumption that these issues in product definition were completely resolved. In the next chapter, we will see the difficulties presented by the clash between this philosophy of immediate correction and the philosophy of defining a stable mode of operation at the plant.

5.1.2 Acoustics

For the acoustics, the timing chain represented a risk that had been acknowledged by the project management team right from the start. The technical solutions selected—three-cylinder engine; solutions for lightening the powertrain such as a plastic oil pan; and the overall lightness of the car—exacerbated both the generation and the transmission of mechanical noises and vibration. For Frédéric Maniodet, customer requirements engineer and the voice of the customer within the project, the problem lay not in the specifications defined in 2011, but rather on competitors' improvements. Other manufacturers had taken giant leaps in this domain, and the Alliance's car had to match them by 2015. "If the car was one step ahead in terms of comfort and handling, that was no longer true for acoustics." Now a series of modifications was launched to resolve this problem: tooth correction of the gears in the gearbox; new "low-noise" belt in the gearbox; new engine mount to cushion the vibrations from the powertrain; new idling settings; etc.

"To move forward, we needed help from the experts at Corporate," said Ludovic Gouère, who was in charge of powertrain in 2ASDU. "Acoustics is a very difficult domain. Even when noise and vibration can be reproduced during simulations, it's far from straightforward to understand what exactly caused them, in order to eliminate them. It took a year between detecting problems and implementing solutions." Now, the challenge was to carry out these modifications alongside the ramp-up of the initial version, so they could be applied as quickly as possible to production vehicles.

Manufacturing approval, signifying that the cars produced have the required level of quality in order to be sold, was obtained at the end of August, 2015. "None of the team really took a vacation, irrespective of their reporting level," recalls Christine Gelin, who replaced Jean François Vial in September, 2015, as Vehicle Project Manager and aggressively tackled ramp-up control and customer feedback management. "Clearly, there were problems—such as the ergonomics of the glove box, which had to be adapted by applying a sticker. But on the whole, the birth of the car went well. The problem now was to ensure ramp-up by stabilizing the manufacturing processes, and attaining repeatability in order to successfully manage the increase in volumes with quality."

5.2 Project Commando Meets Plant Bureaucracy

It was 8 PM on Thursday. In his office, Gérard Detourbet was finishing up his meetings for the day when Céline Buchard, in charge of procuring prototypes and pre-production, knocked on the door. There was an urgent problem with validating the pre-production batch on the assembly line, she explained. Such ad hoc crisis meetings were an everyday occurrence.

In this case, in order to obtain quality approval for the assembly, a complete batch of 25 cars had to be produced, using the actual assembly process and conforming exactly to the mass-production method. Otherwise, approval could not be given and ramp-up would be delayed. However, the project team planned for one of the cars from this batch to have a new "low-noise" timing belt. But if this modification were to be applied to a single car, the assembly process would no longer be exactly identical to the production, which was specifically required to obtain the approval.

How to resolve this? On one hand, the project faced the pressure of increasing ramp-up. Every day, Sales and Marketing were receiving more orders than the plant could produce. Customers were waiting, and deferring the production validation date was simply unacceptable. On the other hand, the modification was crucial in the eyes of the market and had to be made as quickly as possible. If the cars being produced were not used to validate the solution by a trial run, implementation would be delayed and the program would miss the delivery date for the enhanced version, planned for six months after initial delivery. This too was unthinkable.

With the Engine Project Manager invited to join, the meeting began. Detourbet suggested replacing the proposed engine (including the old belt) with the modified one, carrying a new serial number. The validation of the assembling sequence could go ahead as planned, and the car could undergo the durability test for validating the belt, with, of course, specific traceability for the follow-up process.

However, this solution did not comply with the quality process for validating the batches. It amounted to concealing the fact that the engine was not in line with the initial specifications—and, hence, deceiving the standard process, which had to be focused on cars that complied fully with the specifications provided. If the Quality Manager were asked to accept such an exemption, he would decline. But if it were concealed and he discovered it later, that would be even worse.

Discussions grew heated as the executives wrestled with their dilemma. Finally, the Program Director succeeded in championing his solution. But now he had to make a tough call: whether to try and convince the Quality Manager to accept this exemption, or to implement the process without his knowledge. What happened next is another story.

Underlying such crises, which were frequent during the ramp-up period, lay a conflict between two very different philosophies.

On one side, the later stages of development were characterized by the compulsion to shorten the action loop between detecting problems and developing solutions; applying these modifications; testing them; and finally, applying them to production. In spite of the efforts made with myriad simulations to anticipate problems, a handful remained stubbornly difficult to detect—at least in magnitude, if not in nature. This is particularly true of requirements such as noise reduction, which leads to late modifications during the start-up phase. So the project team set up a process to expedite the validation of a new, quieter belt—a customized process built on a commando philosophy.

On the other side stands the need to validate an industrial system for mass production. Here, the main concern is to ensure the repeatability and stability of a process that manufactures a perfectly controlled quality product every minute at the lowest cost. The purpose of validation is to systematically detect and eliminate deviations from the reference process, whether major or minor. Something as small as an error in numbering a part in the information system carried the same weight as a major physical defect in the vehicle. All things considered, coordinating the convergence of no fewer than 1,300 different parts into a sequence of several thousand basic production and control operations, all of which take place inside a single minute, is nothing short of a miracle. In order to validate repeatability, it was essential to exclude exceptional interventions or "adhocratic" ingenuity, and "blindly" apply the standard processes that had been developed to identify and correct all the loopholes. However, in this particular case, introducing the new "low-noise" belt presented a validation challenge—even if it did resolve a real issue for the project team.

After the meeting, further emails and phone calls informed the Program Director of other, equally critical events elsewhere. One reported that Corporate was trying to impose an Alliance engine for the new derivatives, which would be very costly from the Program Management team's point of view. Another highlighted the delay caused by an impact test of the reinforced version of the car. Some were dealt with the same night; others had to wait until morning.

The added value of the project team during this "firefighting" phase lay in its ability to manage its time, energy, and authority to resolve problems based on a deep understanding of the distinct focus of the project. This original purpose had to be restated repeatedly when dealing with stakeholders who arrived with no knowledge of previous compromises and little appreciation of the reasons why "hardwired" corporate responses had been defied.

5.3 Leveling the Competitiveness of the Plant

The launch of production of a new model is always a time of conflict, marked by the collision of contrasting philosophies: the project's (exploration) and the operation's (exploitation). This is even truer when the plant must absorb a new product and, at the same time, make a major rationalization effort—as was the case for the Alliance plant in Chennai. The advent of the 2ASDU project would be an opportunity to regain full control over the plant—and this was enabled by replacing practically the entire hierarchy, from the director to the main managers.

The new Plant Director, Colin McDonald, was a Briton who had previously managed a Nissan plant in Sunderland, UK. Speaking in February 2016, he painted a picture of a facility enjoying a renaissance. "This plant opened in 2010," he explained. "It's the first and only Alliance plant owned jointly by Renault and Nissan—even though, in reality, it's been operated by Nissan until now. Over five years, the plant has launched 25 different vehicles for the three brands [Datsun, Nissan, and Renault]. Lots of products, but they've rarely lived up to expectations. Major launches, and major disappointments in an Indian market that's very demanding in terms of value for money. In this respect, the plant couldn't become a steady-state, high-performing system; it was constantly starting up or downgrading operations because of the weak volumes. You can't rationalize an industrial system in those conditions."

A plant visit revealed a production line that was a world away from the clichés of manufacturing excellence. For example, the end of the assembly line was congested with cars in various states of rework, but this was merely the feeder chamber for another "rework area" located in a different building.

How can we account for such mediocrity in manufacturing performance when Japanese manufacturing was noted for its excellence? Several factors were intertwined. On one hand, as McDonald noted, there was the history of the plant, shaped by a multitude of successive launches that had never delivered the expected stable production volumes. And as the story of Ford shows, the true power of automotive rationalization can only be released by high, stable production rates.

On the other hand, we must bear in mind the limitations of the industrialization approach popularized under the term "transplant." Japanese carmakers tried to use this approach to clone their plants in the various local contexts where they set up. However, this did not work in India. "In Japan, when the manager says something, the subordinates do it; there's no need to check or inspect. Here, they say 'yes,' but then nothing happens." Detourbet says. The Indian working environment was very different from that of a Japanese plant. Staff were poorly trained, if at all, and very unreliable: 40% of the workforce comprised temporary workers hired for the day. The organization of the plant was surprising too, and quite unlike that of Western plants, Renault's especially. For decades, "lean" philosophies have emphasized the overall responsibility of the production line, which integrates various performance elements representing direct work, reliability, and the quality of the facilities. In India, the structure was compartmentalized into a hierarchical production line concerned with only the direct workforce, and centralized, independent maintenance or control departments.

With the Kwid came a chance to turn over a new leaf. "We've got an adapted, modern, competitive product here. At the same time, we must ensure a quick ramp-up to fulfill customer orders, which are growing rapidly, and simultaneously streamline the process," stressed McDonald. However, getting a plant back on its feet while also launching a new product was far from easy, as we have seen; between establishing a new "lean" production philosophy and pragmatically resolving urgent start-up problems, conflicts were common. This was particularly so because the start-up came about within a context of impressive commercial success: it was certainly great news on the whole, but for the plant it meant the constant challenge of prioritizing production within a short period while also optimizing basic processes.

Chapter Six

The Commercial Launch of Kwid in India: March 2015–June 2016

Like the rest of the program, the commercial launch can be thought of as the crossing of three paths. The origins of the first path lie in the history of the Entry program, which became a key element of Renault's international development and an important brand in the emerging markets of countries such as Brazil and Russia. The origins of the second path can be traced back to a study conducted by Renault to set up and operate in India. The third was the one that flowed from the product that was eventually developed and that went into direct competition with the leading brands and their most popular products, such as Maruto Suzuki's Alto and Hyundai's Eon.

6.1 Lessons Learned from Entry

The Entry experience had shown that a dealer network had to be created (or re-created) at the same time as the product hit the market. This created a Catch-22 situation: big volume could not be achieved without a suitable dealer network to reach customers, but it was impossible to persuade investors to back a brand that could not promise high volumes. As a result, Renault's Indian sales teams had to convince dealers that Renault was about to join the exclusive club of car manufacturers who understood the Indian market, and that, in the Kwid, they had a genuine game-changer for India.

According to its sales managers and dealers, Renault achieved remarkable success when it adopted this strategy. This was best illustrated by the success of the Duster, which was manufactured in Chennai and sold in India by Renault under its original name, and by Nissan under the name Terrano. Launched in July, 2012, the Duster was named Indian Car of the Year 2013 and won numerous prizes, building confidence in the brand. In March, 2014, Renault revealed that more than 100,000 vehicles had been sold in less than three years. This proved to be an effective way to communicate with the public as well as investors and was the foundation on which Renault built its dealer network. From only 14 representatives in 2012, Renault had expanded to 200 sales outlets by 2015, and when the Kwid was presented to the dealer network at the end of 2016, the company had 270 sales outlets.

The Duster was thus the key to making a previously little-known brand more popular in the Indian car market. However, it positioned Renault as a high-end brand and meant that the image of Renault became inherently linked to the Duster. The rest of the Indian range consisted of Nissan or Samsung vehicles rebadged as Renault under the names Pulse, Scala, Fluence, and Koleos, but these were only sold in very small numbers. When an Indian version of the Lodgy was launched, it was intended to compete with an Indian bestseller, the Toyota Innova, and to address this problem, but it was also priced at the upper end of the market. The sales price of the two Entry products, Duster and Lodgy, was more than triple that of the Kwid, starting from 8.5 lakhs (€11,000) vs. 2.7 lakhs (€3,500). Competitor analyses conducted throughout the project's life consistently highlighted the lack of brand recognition.

Despite these obstacles, Renault steadily built itself up as an Indian brand. Renault had worked hard to understand India's society and auto industry before attempting to place itself at the heart of the Indian market with what the designers called "Indian 4x4s," by which they meant vehicles shorter than 4 meters that cost less than 4 lakhs (i.e., 400,000 Indian rupees and around €5,300). Proof of this lay in the huge effort that had been put into "Indianizing" the Duster and the Lodgy. The Kwid subsequently benefited greatly from this initiative, and it was the job of the sales and marketing team to accelerate it by reinforcing the image of a reliable challenger. This was of huge importance when dealing with investors who, once committed, would have to shift funds, top sites, or leading dealers to the various brands they represented. The Entry teams had faced similar situations in Brazil and Russia a few years previously.

The duo behind this effort were Sumit Sawhney, who was appointed by Renault in 2012 and became its head of Indian operations in 2014, and Raphaël Treguer, a sales and marketing VP who arrived in September, 2014, from France, where he had been Sales and Marketing Director of Dacia. Having worked with GM and Fiat, Sawhney knew the heads of the distribution groups as well as Indian journalists. Treguer, meanwhile, had a good grasp of the methods developed over more than a decade by the Entry teams to cut marketing costs to the bone. These included doing everything possible to secure press coverage rather than splurging on expensive advertising campaigns. Both knew that they had to communicate directly with the dealer network and the customer at the same time. This strategy had been deployed for the Logan and Sandero in Russia and Brazil and for the Duster in India. The Kwid could develop this practice even more and allow Renault to "shift gears" and "catch up to Maruti and Hyundai."

To make that happen, the sales and marketing team would mastermind something similar to what had happened—without Renault management's consent—during the launch of the Logan. They ensured that the Kwid was a media sensation and not just another new product launch targeted at sales and marketing experts. The team organized a teaser campaign to generate buzz and interest and an influx of orders before the product was even available for test drive. The launch was not only a success with buyers, it also facilitated

Figure 6.1 Presenting the Kwid in May, 2015. From right to left: Gérard Detourbet, General Director of 2ASDU; Sumit Sawhney, Director General of Renault India; Carlos Ghosn, CEO of the Renault-Nissan Alliance; and Laurens van den Acker, Designer Director at Renault (© Renault, reprinted with permission.)

the recruitment of new dealers while reassuring those already in place. In addition, it served to compensate for the shortcomings of the existing dealer network by promoting the product beyond the showroom (see Figure 6.1 on previous page).

6.2 Indianizing the Business

This general strategy would require a strong and specific desire not just to harmonize with the Indian setup, but also to project the impression of being a trailblazer in India and around the world.

With this mindset, Renault launched a Kwid app for Android and iOS that allowed the customer to get to know the product, see it in action, configure it, leave comments, and even reserve the car. Apart from gathering the details of prospects who could be re-contacted later, the app created buzz around the product at very little cost. It also spoke to one of the car's key points of differentiation: "My connection to modernity." Along similar lines, a virtual showroom was developed that allowed internet users to view the model remotely, explore it, and ask questions. This completed a real-world initiative that was based around showcasing the vehicle in busy locations. As sales had started while the vehicles on show were not yet production vehicles, it was illegal to let people get into the car. This limitation was transformed into an opportunity by designing an "interactive display" for this purpose; it created, at very little cost, its own unique "connected" and modern niche. Through such innovations, the marketing team got the press talking about the product, the brand, and the cachet it conferred on India.

This strategy also placed the brand on a par with other leading players by showing Indian professionals and society at large that Renault knew them and understood their requirements. "It's a huge mistake to think that in India, just because there is a demand for cheap products, consumers will be satisfied with the basics," points out Raphaël Treguer. "In fact, the opposite is true—the Indian consumer is extremely demanding and sensitive to innovation. Commercial activism is stronger here than anywhere else in the world. In India, keeping the product range fresh is vital." As illustrated in Figure 6.2, a range of options were made available.

For Renault, constant renewal meant adding successive innovations to a model to remind customers of its existence and to provide new opportunities to talk about it and, it was hoped, buy it. From this perspective, the dealer network had complained about the quality of recent updates to the Duster, which could explain disappointing 2015 sales. Consequently, the approach adopted for the Kwid placated dealers from day one, with assurances that both a more powerful

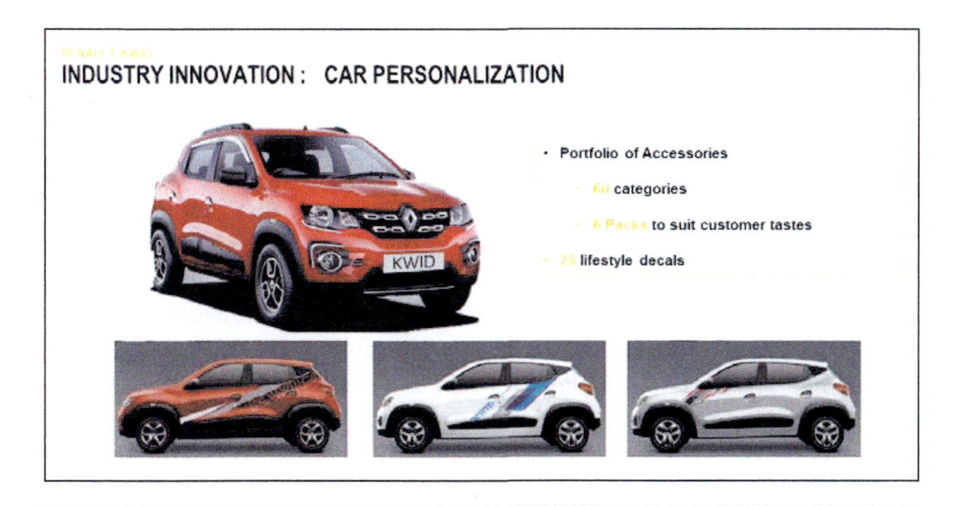

Figure 6.2 The diversified accessory and option catalog, a marketing requirement for the Indian market even in the entry-level segment. (© Renault, reprinted with permission.)

version with a 1-liter engine and a newer version with a robotized gearbox (automated manual transmission), new upholstery, and new "looks" were already in the pipeline.

In February, 2016, two Kwid prototypes named Racer (a sport model) and Climber (a model with the 4WD look) had been presented, but doubts still remained about their long-term commercial potential. Less than a year after the launch, Renault showed its keen awareness of the need to renew the range. The Mumbai design office had to develop the two models not just for the Indian market, but also for all the countries that would eventually buy them. Without spending a single rupee on advertising, public interest was sustained and the dealer network found itself needing help to manage the waiting periods generated by the success of the product. This dealer network was subsequently enlisted for its global deployment adventure.

The product was also launched with a comprehensive range of accessories and options that dealers could use to promote the vehicle itself. These were additional sources of profit, and this, in turn, helped Renault to limit the margins on new car sales.

The fractal nature of the innovation on the product was reflected in the business too: a lot was done with very little. "Indian-ness" and acknowledgment of the unique features of the customer and dealer network played a significant role. The "local content" of the value proposition was the main focus of the launch, in the same way that "local content" had been the main focus during procurement and in balancing the economic equation.

6.3 Throwing Down the Gauntlet to Maruti and Hyundai

In India, serious commercial ambition meant tackling the ultra-dominant player (i.e., the one that controlled the market) head on. This was Maruti Suzuki, whose strength lay in the bulk sales of the Alto in the same way that the Volkswagen brand, after the war, was built mainly on the popularity of the Beetle. With its Eon model, Hyundai had proved that it was up to the task. Datsun, facing the same challenge, had responded with the GO, which failed to make an impression. Renault thus had to prove that, even though it was a part of the same Alliance and shared the same plant, it could do differently—and better. At the same time, it had to convince the dealer network—and new investors—that brand recognition gained through the sales of the Duster, while still modest, would allow it to achieve what only Hyundai had so far managed to do: establish itself as a foreign car manufacturer.

The main strategy—as it had been for the Logan and then the Sandero—involved relying on the product and its ability to demonstrate its superiority when compared to the competitors' products. The product itself would function as a kind of warranty of Renault's expertise, involvement, and understanding of India, the Indian market, and the constraints and interests of its dealers. Just as Carlos Ghosn had been convinced by the LCI's short film, from the first mention of the Kwid until the first orders were placed and the first sales were made, the product and its key sales arguments were consistently highlighted. It was important to lend credibility to this core objective even before the car itself was visible on Indian roads. To this end, the sales and marketing team explored every possible avenue "to create interest before the launch" across all available media and potential points of contact (see Figure 6.3). The media interest generated was widely shared and particularly strong on social networks and was, to some extent, imposed on the dealer networks and sellers. They felt that they had embarked on a *success story* that largely echoed Renault's four key points of differentiation for the product.

Renault had two major handicaps: a poor brand image and a dealer network with very low reach. Its target audience included many first-time car buyers, as well as buyers who were particularly interested in how easy it would be to get their car serviced and repaired. Renault's innovation was to transform this weakness into a strength by turning the product's relatively small physical presence into a launch strategy.

In order to create a buzz about the "Renault innovates" concept—and to respond to customers' objections about the distance (and travel time) to their nearest Renault dealership—the sales and marketing team introduced the concept of a mobile workshop called the "Workshop on Wheels," which was designed to

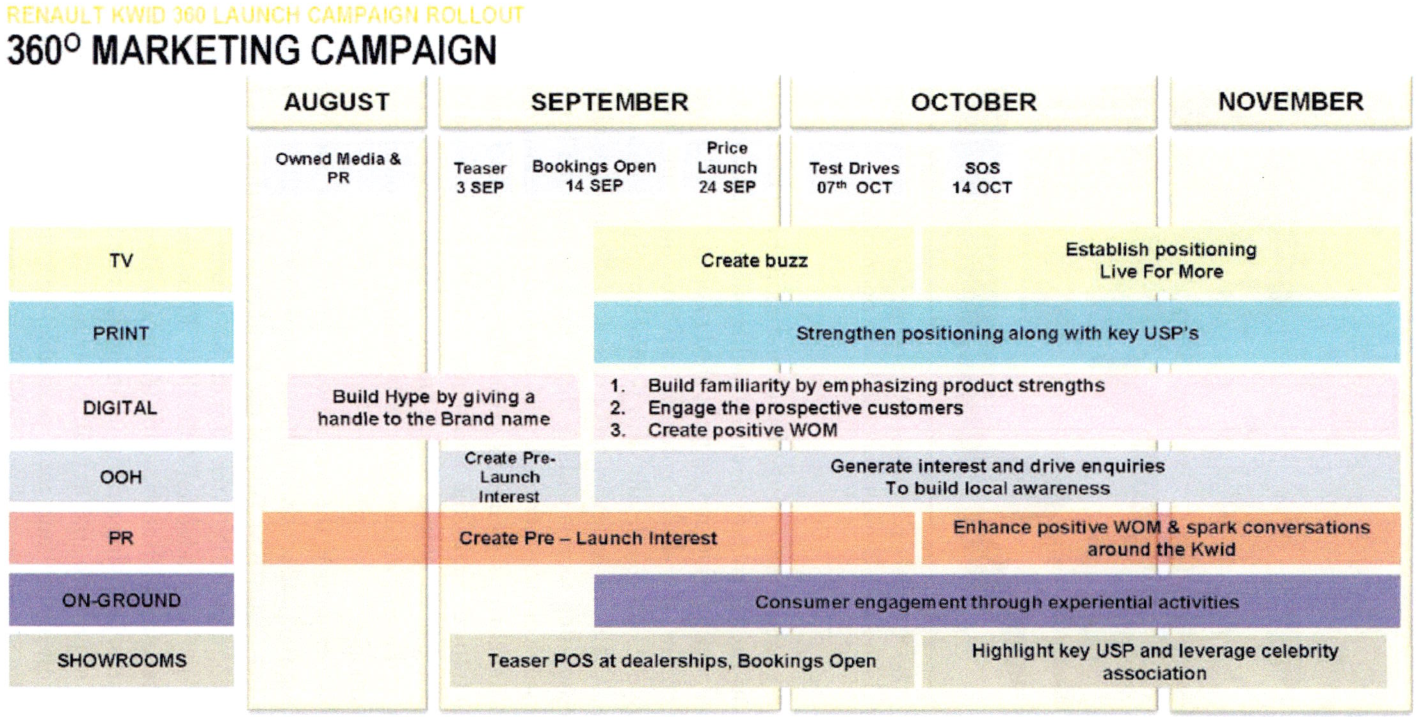

RENAULT KWID 360 LAUNCH CAMPAIGN ROLLOUT

360° MARKETING CAMPAIGN

AUGUST
SEPTEMBER
OCTOBER
NOVEMBER

Owned Media & PR
Teaser 3 SEP
Bookings Open 14 SEP
Price Launch 24 SEP
Test Drives 07th OCT
SOS 14 OCT

TV
Create buzz
Establish positioning Live For More

PRINT
Strengthen positioning along with key USP's

DIGITAL
Build Hype by giving a handle to the Brand name
1. Build familiarity by emphasizing product strengths
2. Engage the prospective customers
3. Create positive WOM

OOH
Create Pre-Launch Interest
Generate interest and drive enquiries To build local awareness

PR
Create Pre – Launch Interest
Enhance positive WOM & spark conversations around the Kwid

ON-GROUND
Consumer engagement through experiential activities

SHOWROOMS
Teaser POS at dealerships, Bookings Open
Highlight key USP and leverage celebrity association

Figure 6.3 The Kwid marketing plan.

Figure 6.4 The "Workshop on Wheels" concept, an answer to the small dealership network in India. (© Renault, reprinted with permission.)

ensure the regular presence of the brand and its technical teams in particular areas or towns (see Figure 6.4). Far from taking offence at this, dealers representing the brand welcomed it, and many applied to open new sales outlets.

Collaborations with dealer networks and promotions in shopping centers yielded prospects' contact details. This pool of prospective customers was reassuring to have but, in reality, it was never actually used, as the launch was already a resounding success without them. Even before sales began and when new dealers still had to be recruited, sales and marketing became more demanding in terms of existing sales outlets, requesting more sales staff and insisting on having them trained. Again, the consistency with the product strategy was quite clear. Marketing an entry-level vehicle and focusing on costs were no reason to cut corners, and managing customer relationships during ordering and delivery was no exception.

From this perspective, the launch was a great success and played a major role in placing the product on an almost equal footing with the Alto within a few months (see Figure 6.5). Sales of the Kwid exceeded those of the Hyundai Eon from December, 2015; after 10 months, almost 80,000 Kwids had been delivered, in spite of production difficulties. Firm orders placed over the first 11 months were quite high at about 165,000 units. The delivery time announced was four months, justifying a ramp-up in production beginning in May, 2016. The goal was to produce 10,000 vehicles per month, most of which would feature the highest-quality (and most profitable) specifications. Four configurations of the Kwid were available: E0/STD corresponded to the lowest price,

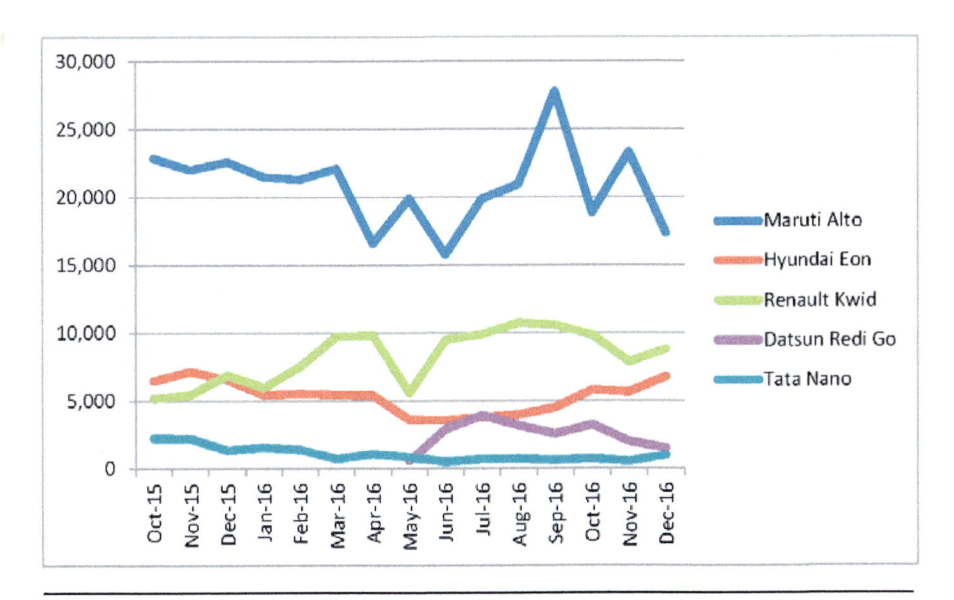

Figure 6.5 Sales comparison for Alto, Kwid, Eon, Redi-GO, and Nano for the first 14 months after launch. (Source: TeamBHP.com. June 2016: Indian Car Sales Figures & Analysis. www.team-bhp.com/forum/indian-car-scene/177998-november-2016-indian-car-sales-figures-analysis.html.)

with various levels of added features characterizing the more expensive RXE, RXL, and RXT (featuring built-in satnav). Since launch, 80% of the orders were for the top-end RXT model, compared to the forecast level of just 60%. Additional options and accessories mean bigger margins, and customers are paying for such extra features when they buy the Kwid.

6.4 Success Comes at a Price

This partly unanticipated success created an urgent problem, relatively unusual at this level in India: customer waiting times. On one hand, sales had surpassed expectations, and on the other, the production ramp-up had been more difficult than expected. From the end of 2015, the dealer network blamed these problems, along with missed deadlines for delivery times, for lost and canceled orders. Dealer networks claimed that Renault was at fault for breaking its commitments and/or prioritizing other sales outlets. The manufacturer, however, accused some dealers of giving unrealistic deadlines in order to win over customers.

As the issue grew, the business found itself, paradoxically, expending its energy on managing the fallout from a launch that had been almost too successful for its own good. As a result, development objectives outside India had to be slightly curbed. The situation was exacerbated in June, 2016, when the Redi-GO—the second Alliance model developed by the 2ASDU team—was launched on the same platform. Assembled on the same line, it had obviously absorbed part of the capacity and energy that the teams could devote to managing the production ramp-up. As a result, it was only in the summer of 2016 that a production rate was achieved to meet actual demand at around 10,000 units per month.

It was in this context that in May, 2016, Global NCAP organized an impact test of five Indian vehicles: the Kwid (in its November version), the Maruti Suzuki Celerio, the Maruti Suzuki Eeco, the Mahindra Scorpio, and the Hyundai Eon. None won a single star for the safety of adults seated in front. This was a disappointment for Renault, particularly as the version tested had already been modified to improve security, and the engineers had hoped for better results: "We came so close to winning a star: within 3 mm, in fact . . ."

This result was the only cloud on the horizon. To avoid further disappointments and ensure the product's future in contexts where standards were stricter (such as Brazil, where the Latin NCAP applies), a new version was developed that was guaranteed to get at least one star, and it entered production on May 9, 2016. Sumit Sawhney emphasized that subsequent tests of this new version would make up for previous poor results and continued to maintain a momentum in the media by insisting that the Kwid offered customers more for the same price.

As is often the case with emission results in Europe, the evidence showed that such tests were less important for the public than for the press and for carmakers' relationships with political and administrative circles. Even for this second group, reactions differed sharply within India and outside it. "We waited for the NCAP results to spark an important debate in India, but it didn't happen," recalls Gérard Detourbet. "But it was almost a non-event here." In addition, in Brazil and Algeria, where interest in the Kwid was strong, commentary on the news of these results implied that India was a special case. Selling the Kwid in India was crucial for the program—but success there did not guarantee the same elsewhere. The Brazilian launch needed to be completely unlike the Indian one, as the standardization for Brazil (assembly and supplying) would be considerably different from that of India.

The sales effort for the Kwid had to be adapted to the local context: to gain economies of scale and remain "low cost," firms would have to sell a product that is essentially the same everywhere and highlight the same key sales arguments in every market. At the same time, in the low-end market more than in any other, the car should appear Indian in India and Brazilian in Brazil. Renault had to show society, the media, dealers, and customers that, like a Maruti Suzuki in India or a Fiat in Brazil, it understood that one size did not fit all. It also knew that its ongoing efforts in this regard were signals of a lasting commitment that was worthy of support.

Selling such a product first meant achieving large volumes and increasing market share, but this depended on developing the dealer network. The sales and marketing team needed to sell both the car and the development program to investors and customers. The successful development in India in 2015–2016 showed that this could be achieved with careful work in the field, approaching potential investors one by one, alongside a national multichannel marketing campaign. This local effort sparked a more global wave of interest. In India, Sumit Sawhney and Raphaël Tréguer began to consider how that wave could be harnessed for the launches to come. They might be using the same ingredients, but they would be preparing quite a different dish.

Chapter Seven

Prospects for International Deployment of a New Lineage: 2016 and Beyond

In our book *The Logan Epic*,[1] we showed how the economic success of the Entry program was due not only to a clean and early break with the past (the X90/Logan project), but also to the ability to build on this beginning with a series of new developments, expanding both the product scope of the Entry range and the geographic scope of the markets covered.

We emphasized that this deployment capability was based on the collective expertise of the firm, which we call *project lineage management*.[2,3] Whereas regular project management optimizes a design on a defined timeline in order to complete the project, project lineage management takes on the more challenging task of uniting three disparate, wider-ranging strategies: building

[1] Julien, Bernard, Yannik Lung, and Christophe Midler. *The Logan Epic: New Trajectories for Innovation*. Paris: Dunod, 2013.

[2] Midler, Christophe. "Implementing a Low-End Disruption Strategy Through Multi-Project Lineage Management: The Logan Case." *Project Management Journal* 44, no. 5 (2013): 24–35.

[3] Maniak, Rémi, and Christophe Midler. "Multiproject Lineage Management: Bridging Project Management and Design-Based Innovation Strategy." *International Journal of Project Management* 32, no. 7 (2014): 1146–1156.

knowledge and continuity to preserve the "DNA" of the initial project; learning and entrepreneurship to seize new opportunities and "bounce back" by developing products adapted to new target conditions; and rescaling the organization via the transition from a co-local "commando" project team to a globalized program management, while delicately balancing the consistency and diversity of various developments.

This deployment approach, which combines the engineering and marketing efforts of multinationals in a unique way,[4,5] historically represents a new stage in the general rationalization of innovation processes of major groups in the 2000s. It breaks with the more traditional strategy of using standardization/derivation of ethnocentric platforms to redefine products from the groups' home markets.

Just as we were completing our book, there was a clear shift toward project lineage management. It is obviously too early to predict its results; moreover, confidentiality prevented us from revealing any plans. Nevertheless, it is possible to analyze the new horizons that might be opened up, as well as the foreseeable difficulties that would have to be overcome in order to explore them.

In comparison with the Logan, the Kwid's radical break with the past is expanding deployment potential while also raising issues with its implementation. Specifically, the radical "Indianization" of the project would be crucial to achieving its ambitious targets—and, in turn, would raise the problem of adapting the product and processes to utterly different production and market conditions.

7.1 A Difficult Transition: The End of a Project and the Birth of a New Lineage

Before considering these different development potentials, let us return to the team we left in Chennai. The years 2015–2016 were a period of change for the 2ASDU team. The sales of the Kwid in India brought the first chapter to a definite close. Now, the task was to create a product line around the first model.

In June 2016, the Datsun-branded product, the Redi-GO, was launched. The Kwid Brazilian launch was already scheduled, and a request had already been made for an operation to fine-tune the product to the specific expectations

[4] Pechmann, Felix von, Christophe Midler, Rémi Maniak, and Florence Charue-Duboc. "Managing Systemic and Disruptive Innovation: Lessons from the Renault Zero Emission Initiative." *Industrial and Corporate Change* 24, no. 3 (2015): 677–695.

[5] Ben Mahmoud-Jouini, Sihem, Thierry Burger-Helchen, Florence Charue-Duboc, and Yves Doz. "Global Organization of Innovation Process." *Management International* 19, no. 4 (2015): 112–120.

of this market. In terms of engineering, this meant optimizing the industrialization, initially considered in an Indian context, for a globalized context. In the Indian market, as we have seen, there was no question of letting the Kwid's initial success run out of steam: new versions had to be quickly rolled out and a buzz created every six months by launching options that would make the brand the talk of the town. Consequently, far from being the end of 2ASDU activity, this year witnessed a convergence between the launch vehicle's ramp-up phase and the lines of development that would extend it.

However, this continuity in development had to bridge a discontinuity within the organization. Because the normal assignment duration within groups such as Renault or Nissan is four years, the first expatriates of 2012 left the project in 2016; the product launch constituted a natural milestone. The way forward to renew the team was far from clear: the task going forward was less attractive, and engineering rationalization in the two parent companies had depleted recruitment pools. Moreover, it was hard to prevent Indian engineers with such a glowing success story on their CVs from flocking to competitors—a classic problem. Finally, the HR management standards of RNTBCI corporate, which set salaries, did not allow for competitive remuneration. As a result of all these factors, far from being plain sailing, this stage saw earlier issues and negotiations resurface.

7.2 Kwid's Indian Product Lineage

As had been the case for the Logan, the first expansion of the Kwid program added a range of product innovations for the Indian market. As mentioned, the aim of the strategy was to extend the "wow" factor of the Kwid launch by introducing innovations with USPs as quickly as possible. "We don't want a repeat of the Duster episode, when we didn't know how to keep sales alive throughout the vehicle's lifetime," said Detourbet. "This time round, we want to stop competitors catching up by offering innovations every year." From the autumn of 2016, the automated manual transmission and one-liter engine would be available. Other products were already in the pipeline. All included features that were functionally important for the low-end Indian market and also offered strong differentiation.

At the end of summer 2016, a new variant of the Kwid had been proposed with a more powerful engine of 1 liter (the first engine available was 0.8 liters) for an additional price of only 22,000 rupees (€300), extending the price range of the car up to 4 lakhs (€5,300). With this "Kwid 1.0," Renault can be more competitive on the Indian market, offering an alternative to competitors' cars, such as the Maruti Alto K10.

The automatic transmission (AMT) was a functional feature that had just been introduced by competitors. Using Magnetti-Marelli technology, the Alto offered it in September, 2015; Tata offered it for the Nano in May, 2015; Hyundai, like Renault, monitored these movements but would not be ready to offer the feature until 2018. The robotized gearbox is very attractive to drivers who are accustomed to driving two-wheelers and are therefore unused to changing gears. Indian traffic conditions made automatic cars even more attractive, with drivability improved by functionality such as throttle kickdown. So while the early introduction of this option might have seemed out of place for a low-end vehicle in Europe, it was essential in terms of use value in the Indian market.

All that remained was to offer the option for a price that, as usual, maintained a meaningful price competitiveness threshold. The automatic variant had been planned from the start, with a structural design option that facilitated, on the one hand, the addition of automation, and on the other hand, the development of an ECU (electronic control unit) integrating the engine and gearbox control. Moreover, the same supervisor, Michel Anastasiou, designed both the manual and automatic gearboxes.

All this indicated that, just as the Logan had welcomed a little sister followed by two brothers and another sister three years after its launch, the Kwid and Redi-GO twins would soon have their own siblings, signaling the start of a family.

7.3 International Expansion Plans Confirmed

Carlos Ghosn's orders were clear: the Alliance had to capture 10% of the global car market, and this was to be done through an aggressive policy for emerging markets, starting with the famous "BRICs" (Brazil, Russia, India, China). Thus, when the option of designing a new vehicle that introduced the CMF-A "family" had been taken, the project was scheduled to start in India from 2015, but deployment in other markets had to follow quickly. Therefore, from the start, the goal had been to design a global product initially targeting India, Brazil, and Russia for an annual volume ranging from 420,000 to 530,000 XBA/I2 vehicles (2012). China was not included in the initial projections, but Renault had planned other locations for a volume of more than 800,000 cars per year for the Alliance.

Program Director Gérard Detourbet, who had pushed through the intercontinental deployment of the Entry range, invested heavily in the internationalization of the new program. Drawing on his knowledge of various industrial sites and markets, he traveled the world to plan this deployment and convince the Alliance, especially Renault, to make a commitment. 2ASDU was confirmed as the project manager of this internationalization.

In 2013, the launch dates in India (summer 2015) and Brazil (autumn 2016) were confirmed (thus, they would be staggered), but the collapse of the Russian car market led to the postponement of the car's introduction until the end of 2018. In July, 2014, the XBA's international mission was still set fair for Renault, with a planned annual volume in the range of 565,000 to 615,000 vehicles, while the objectives set out for Datsun's Redi-GO remained more limited: only Indian or neighboring markets (Nepal, Sri Lanka, Bhutan, Bangladesh) were confirmed, even though other locations were still being explored.

Two years later (summer 2016), nothing had changed, and decisions made did not allow for this international deployment to be accelerated—even though, at the end of 2015, the Program Director had spoken to the press about an annual volume target in the region of one million vehicles based on the CMF-A platform over time. This announcement revealed that various ideas were still being considered. As with the Entry program, some avenues seemed to be closed off, at least for the moment, while others were reopened. So the European press widely covered the possible launch of the Kwid in Europe. Its production would be carried out in Tangier, where the Alliance plant had overcapacities, but Renault did not immediately appear to confirm these rumors. It should be recalled that the Logan would not be aimed at the European market, where the Entry range had finally occupied its rightful place. Finally, at the beginning of 2016, Carlos Ghosn's announcement of a *low-cost* future Renault electric vehicle in China fueled speculation about a possible electric version from the CMF-A platform as a possible base. For now, the focus was on the success of the Brazilian launch of the Kwid, planned for the coming months.

7.4 Brazil: The First International Deployment

In 2013, when the decision was made to produce the Kwid in the Renault plant in Curitiba, the economic situation was highly favorable. Brazilian car sales exceeded three million units for the first time, and Renault's Entry range (Logan, Sandero, Duster) had its fair share. Nissan opened its own plant in Resende. The only cloud on the horizon was that Renault did not have a car that could compete on the *"carros populares"* market (sub-one-liter engines); its candidate was a Clio 2, which had much less appeal. However, this segment represented almost 40% of total market sales and, apart from the shortage of volume that always left the Alliance behind the leading quartet of VW-GM-Fiat-Ford, it was a handicap for Renault because the average fuel consumption of its range was the highest compared to that of its direct competitors. Yet the new INOVAR-AUTO regulation adopted by the Brazilian government at the end of 2012 set a target with regard to energy consumption that could penalize Renault financially unless it made improvements—and the Kwid could deliver them.

In the summer of 2016, Renault was busy in Curitiba beginning mass production before the São Paulo Motor Show in November, 2016, in order to sell the Kwid at the beginning of 2017. But the situation was altered by a sudden market downturn: economic growth stalled, sales dropped, Brazil's national currency (the *real,* R$) weakened against foreign currencies, etc. From 2012 to 2016, car sales declined from 3.1 million to less than 1.7 million. All in all, sales volume was halved compared to the peak of 2012–2013. However, the structure of the market had also developed, because Brazilian customers who were still active in the new car market showed a preference for better-equipped models, just as their European counterparts did. There was a decline in new car sales: thus, the one-liter car segment, which had peaked at 70% sales in 2001, amounted to just 34% in 2015. Cars costing less than R$35,000 could no longer find buyers, and car manufacturers moved upmarket by withdrawing their most basic models to offer better-equipped vehicles whose basic price hovered around R$40,000. For now, the Fiat Mobi, Ford Ka, Volkswagen Up!, and Chevrolet Onix were Renault's competitors, and it used the Sandero to compete with them while waiting for the Kwid to be rolled out.

In Brazil, Renault no longer had to compete with Hyundai's Eon, but with its HB20, a vehicle designed for the Latin American market and produced in Brazil from 2012. In a few years, the Korean car manufacturer managed to sell more cars than Ford and move closer to the sales of Fiat with a reduced range comprising two cars. Therefore, the Kwid would face a difficult challenge when launched in the Brazilian market: in the summer of 2016, annual sales forecasts were halved in comparison to initial expectations (100,000 cars sold annually) in the hope of an upturn in future sales, and there were still ongoing issues regarding the positioning of the car's list price.

However, before the car could be manufactured and sold, it had to be adapted to the South American market, production flows between the plants had to be organized, and procurement had to be reconsidered. The Brazilian branch of 2ASDU took responsibility for this. In Brazil, 2ASDU worked independently from Renault, while still making use of its infrastructure. Gérard Detourbet entrusted Bertrand Ciavaldini, who managed the Renault launch in India from 2008 to 2012, with leading the 2ASDU team in Curitiba, and granted him a large degree of autonomy. In fact, while the Brazilian Kwid (codenamed "XBB") bore a family resemblance to its Indian sister (the "XBA"),[6] it was far from being

[6] The first letter X (XBA) was generic for all types of bodies. The second letter, B, corresponded to the body: a hatchback. The last letter corresponded to the country: A for India, B for Brazil. To be followed for other bodies and countries in the years ahead.

the same car. It took engineering ingenuity to part-redesign the product and prepare it for industrialization. The platform brought together up to 160 people, of whom just over a quarter were 2ASDU, while the rest remained with Renault but were placed under the functional supervision of Bertrand Ciavaldini. There was no Nissan presence because there was no immediate project of launching the Datsun brand (and hence the Redi-GO) in Brazil (even if other vehicles derived from the CMF-A were possible). Just two Indian staff members supported the launch of the project for a year; the project office housed mostly Brazilians, plus fewer than 10 expatriates and some Argentinians and Colombians. The Indian 2ASDU team also made a big contribution to completing the redesign work, because the overview and details of the vehicle were conceived at RNTBCI in Chennai.

The product had to be adapted to meet both local regulations (right-/left-hand drive; ABS safety standards; airbags fitted as standard; side airbags, etc.) as well as market expectations in terms of engine power (1-liter engine only, and flexfuel) and the other preferences of Brazilian customers: 14″ wheels (instead of 13″); colors; analog instruments in place of digital; internal rear-view mirror adjustment; reclassification of internal plastics; etc. In contrast to the usual approach, which involved simplifying (decontenting) a model intended for central markets (Europe), the approach here was to improve the original Indian product (XBA), which had been designed with the greatest frugality: it was enhanced and "taken up a notch." This upgrading approach proved unsettling for the organization—especially the marketing department, which dragged its feet. "I pushed to move upmarket, but it wasn't easy," recalls Bertrand Ciavaldini. Enhancements have been more wide-ranging than originally envisaged—for example, the external rear-view mirror adjustment became electric. This was not always in line with the Program Director's doctrine, but it was managed. Alignment with quality regulations could sometimes be more difficult in Brazil, where standards were closer to European levels, in contrast to less exacting Indian standards. This meant revisiting the problem of engine noise (see Section 5.1.2 on page 63).

To prepare the Kwid, the production capacities of the plant in Curitiba (Brazil) were increased from 280,000 to 390,000 vehicles per year by introducing an additional team and through investments in the assembly plant. The engine plant, which would assemble parts coming from India while manufacturing the heaviest and largest parts (such as the cylinder head) on site, was also upgraded. Staying with the powertrain, the Renault transmission plant in Chile would machine the pinions and assemble the gearboxes using housings imported from India. Finally, to make room for the Kwid, a part of the Entry range that was previously assembled solely in Curitiba for the South American markets (MERCOSUR) was transferred to the Renault site in Cordoba, Argentina, which would produce the Sandero and Logan. All these investments incurred a relatively heavy expenditure for the Kwid, especially as the project was characterized by the re-internalization of the production of certain components and equipment that were usually outsourced.

As far as procurement was concerned, the local content rate of the Kwid would be far removed from the Indian 97.5%. Brazil is thought of as an expensive country, with costs approaching European levels. Labor costs are clearly higher than in India, and the local automotive supplier industry is dominated by the multinational companies that the program had sought to avoid. The answer was a combination of three procurement sources: imports from India, renegotiation with local suppliers, and internalization of certain subassemblies by Renault.

The difference in costs with India was such that, in many cases, it was preferable to import parts and pay high taxes rather than manufacture them on site (even with logistics costs included). This meant that almost half of the 1,300 referenced parts (45%) would be imported from India—both parts shared with the Indian version (XBA) and those produced only for the XBB (40%). Just over a quarter of the parts would be produced in Brazil—especially large parts with high added value and, usually, components for which the transport costs would be exorbitant. Overall, the local content rate would be 45% at the outset, gradually increasing to 60%.

In this local procurement, the design-to-cost idea was extended further. As the purchasing team was 2ASDU and not Renault, supplier negotiation was based on different foundations, sometimes choosing suppliers who could be rejected from the Alliance panel in Brazil. This, of course, included certain multinationals in Brazil that could not be avoided, but 2ASDU also encouraged certain medium-sized local companies to become suppliers.

An example that illustrated this approach was the seat assembly process. The OEM that traditionally supplied the Curitiba plant with seats for various Renault models did not offer competitive prices. As had happened in Tangier, the decision was made to assemble the seats internally by creating a space within the Renault plant, using components from Brazil as well as India. In this way, Renault encouraged the retention of a Brazilian producer for the seat foam to avoid time-consuming and costly transportation from India—the seat covers would be cut and seamed by a local company which made jeans, and which was given support to handle this new line of business; the metal seat frames would be imported from India.

Another example of the Indo-Brazilian mix was the dashboard (IP). It was designed in India by Motherson, which also prepared the molds. These had been validated on site before being sent to Brazil, where Renault had also decided to take responsibility for the manufacture of dashboards by investing in the plastics manufacturing equipment that was used for other components as well. All in all, around 16% of the referenced parts were to be produced internally.

In the end, the cost price target had to be met—even if that cost was less than around 30% of the cost of the Sandero in Brazil, the product was enhanced, and local production costs were higher. Even so, the final product was a lot more expensive than in India—almost double, in fact (80% higher at the prevailing rate of exchange). While the presentation of the car to the press generated positive reviews, at the time of writing (December, 2016) it is clearly too early to measure how well the Kwid will be received by the Brazilian market. However, the story so far offers lessons that can shape the future trajectory of international deployment in areas such as range enhancement, based on the product-line management strategy that we previously observed on the Entry program. By ingeniously linking the capitalization of learning with adaptability, this strategy prepared this charming little car to take the opportunities it would no doubt discover on its travels.

Part II

Fractal Innovation, Frugal Engineering, and Emerging Countries Growth Strategy

Frugal innovation is very much in vogue, having spread like wildfire in both the academic world and professional circles.[1] However, that is not to downplay the importance of the concept merely because of its popularity: we have long known that the mindset behind managerial fads[2,3] generally follows underlying trends. It has driven important shifts in management doctrines, prompted managers to ask useful questions, and helped to shift possibly outdated organizational orthodoxies inherited from corporate history. However, such phenomena also have their limits and downsides. First, the often simplistic discourse has the advantage of being persuasive and easy to communicate, but it becomes a source of frustration once it shifts from inspiring sentiments to hands-on implementation.

---

1 Tiwari, Rajnish, Louise Fischer, and Katherina Kalogerakis. "Frugal Innovation in Scholarly and Social Discourse: An Assessment of Trends and Potential Societal Implications." Joint working paper of Fraunhofer MOEZ Leipzig and Hamburg University of Technology in the BMBF-ITA project, Leipzig and Hamburg, 2016.

2 Midler, Christophe. "Logique de la mode managériale." *Gérer et Comprendre,* no. 3 (1986): 74–85.

3 Abrahamson, Eric. "Managerial Fads and Fashions: The Diffusion and Rejection of Innovations." *The Academy of Management Review* 16, no. 3 (1991): 586–612.

Also, fads are fleeting—indeed, that is what makes them fads—whereas organizational transformations take much longer. A fad has the power to unite both the focus of the managers and the energy of the employees at the same time, and to mobilize resources and time. However, it would soon be supplanted by another fad, which would make it harder to continue down the same path—if it did not invalidate it altogether. Finally, while fad discourse is very wide ranging, the validity of the underlying theories is uneven: what was true for the automotive companies does not necessarily hold for companies with very diverse markets, such as Air Liquide or 3M, and *a fortiori* for an SME.

Faced with these problems, researchers have two options for overcoming the fascination (or rejection, for that matter) of managerial fads. First, they can question the new doctrine in the light of theories produced by the management discipline. Beyond the fanfares of innovation and breakthrough that drove the fad to hard-won media prominence, what issues would it need to address in order to support (or oppose) the current body of theories? Academics can give weight to this process first by revealing the fad's deeper theoretical resonance, beyond the innovation rhetoric; second, by questioning it in the light of in-depth analyses of its implementation; and finally, by exploring the question of "how"—whereas popular doctrines are generally limited to "why."

With this in mind, the in-depth analysis we have presented here offers many advantages. It illustrates the possibility of applying much-vaunted management doctrines by showing how they can be combined with the reality of a pre-existing organization and accommodated with the other imperatives and mindsets that are inevitably promoted by a large company. This second part of the book explores these perspectives by focusing on the three key dimensions of the Kwid:

- First and foremost, the nature of the innovation that this project pursued, as well as the processes and organizations that could support it.
- Second, the understanding of the strategic process by which—from Louis Schweitzer to Carlos Ghosn, from the Logan project to the Kwid project—a heterodox vision of the search for profit through innovation in the automotive industry could be developed and maintained.
- And finally, the in-depth study of the ongoing deployment of multinational corporations, from home-centric innovation processes to multi-centric globalized innovation processes in which emerging countries appear to be an opportunity for the emergence of frugal innovation, which can then be deployed in a reverse way to mature markets.

Chapter Eight

Fractal Innovation and Creative Product Development

The academic literature addresses the concept of innovation from two different perspectives.

The first—based on the economics of Schumpeter and the literatures of strategic management and marketing—is interested in innovation as the end result. What effect does it have on customer uses, markets, and competitive positions? *Sustaining* or *disruptive* innovations form part of this perspective. Likewise, the term *frugal innovation* describes the properties of the products or services as "doing more with less,"[1] highlighting the relevance of such initiatives in emerging and developing countries, which combine the high solvency constraint of customers with the shortage of bid-system resources and inadequate institutional regulations ("institutional voids"[2]).

We will deal with this perspective in the following chapters. For now, we focus on the other aspect of the innovation literature, which concerns the processes by which the innovations are created within companies—or, more broadly, within

[1] Radjou, Navi, Jaideep C. Prabhu, and Simone Ahuja. *Jugaad Innovation: Think Frugal, Be Flexible, Generate Breakthrough Growth.* San Francisco, CA: Jossey-Bass, 2012.

[2] Khanna, Turan, Krishna G. Palepu, and Jayant Sinha. "Strategies that Fit Emerging Markets." *Harvard Business Review* 83, no. 6 (2005): 63–74.

design systems.[3] Our analysis comprises three stages: Initially, we will deal with the concept of *fractal innovation* to define the nature of the innovative breakthrough of which the Kwid was the archetype. Next, we will specify when and under what conditions it emerged into the design process. Finally, we will analyze the organizational dynamics deployed with regard to the overall organization of the design system.

8.1 Fractal Innovation

Among the standard typologies of innovation, the dichotomy between "breakthrough innovation" and "incremental innovation" is probably the most common. *Breakthrough innovation* is responsible for profoundly destabilizing the design system in place, whereas *incremental innovation* pertains to a strategy based on the continuous, cumulative improvement of a "dominant design"[4,5] within the existing design system. However, the case of the Kwid revealed the ambiguities within this apparent conflict. As a car, the Kwid did not introduce any remarkable breakthrough. However, as a classic automotive product that was modern and attractive, priced at €3,500, and also profitable, there was no doubt that it provided an unprecedented experience in automotive history. Indeed, it was precisely this capacity for matching price and value proposition that created the breakthrough. So how should we classify it?

The typologies proposed by Abernathy, Clark, Henderson, and Utterback in their ground-breaking works articulating the strategic problem of innovation and the engineering sciences (*engineering design*) were of little help. They revealed how important the concept of architecture was, both as a catalyst for innovation in the product and as a destabilizing factor of the bid system.[6,7] Innovation could take place either in the components of the protected-architecture product (modular or component innovation), or through a change in the architecture

[3] Ben Mahmoud-Jouini, Sihem, and Christophe Midler. "Compétition par l'innovation et dynamique des systèmes de conception dans les entreprises françaises ; une comparaison de trois secteurs." *Entreprises et Histoire,* no. 23, 1999.

[4] Anderson, Philip, and Michael L. Tushman. "Technological Discontinuities and Dominant Designs: A Cyclical Model of Technological Change." *Administrative Science Quarterly* 35, no. 4 (1990): 604.

[5] Utterback, James M., and William J. Abernathy. "A Dynamic Model of Process and Product Innovation." *Omega* 3, no. 6 (1975): 639–656.

[6] *Ibid.*

[7] Henderson, Rebecca M., and Kim B. Clark. "Architectural Innovation: The Reconfiguration of Existing Product Technologies and the Failure of Established Firms." *Administrative Science Quarterly* 35, no. 1 (1990): 9.

Table 8.1 Typology of Innovations According to Henderson and Clark[7]

Components / Architecture	Improved	Modified
Unchanged	Incremental innovation	Modular innovation
Changed	Architectural innovation	Radical innovation

itself (architectural innovation), or through a combination of the two (radical innovation). (See Table 8.1) In this sense, the digital camera or Dyson's bagless vacuum cleaner are archetypes of radical innovation.

According to this typology, the Kwid would be placed in the "incremental innovation" box: no new architecture, no component that was radically changed by a new technology or material. However, obviously, such an analysis would oppose the significance of the breakthrough that the Kwid's design brought about—for example, the destabilization of the processes and organization of the engineering teams in place.

Faced with the inadequacy of the existing categories, in Part I we proposed the concept of fractal innovation, which we can now define. Fractal innovation refers to an approach in which the product definition is systematically questioned, along with the full gamut of design variables (product, process, location; and industrial options, suppliers, marketing modes) and at all levels, from the overall sizing of the project to the definition of the characteristics of each component; from cable diameters to the characteristics of screwdrivers on the assembly line.

Fractal innovation also greatly differs from an improvement approach that is spread out, continuous, and incremental—popularly known by its Japanese name *kaizen*.[8] *Kaizen* (often translated as "continuous improvement") begins with the existing process and improves it incrementally over time. Here, fractal innovation is part of the specific breakthrough period that constitutes the development of the new product. Moreover, *kaizen* is basically an approach whereby operators streamline a process by comparing current and past performance. As those who have visited car manufacturing plants know, the *kaizen* approach is put into practice using a range of graphs that are posted in team meeting rooms and that, generally, rise inexorably from past to future performance. In contrast, fractal innovation works toward a completely exogenous

[8] Imai, Masaaki. *Kaizen, the Key to Japan's Competitive Success.* Singapore: McGraw-Hill, 1991.

objective, imposed by the performance of competitors and/or the economic dictates of customers. While *kaizen* progresses step by step, the objective of fractal innovation may seem almost unattainable at the outset. Moreover, in fractal innovation, the existing process is not privileged over other options that may be more likely to achieve the target. While *kaizen* follows a strategy driven by successive bottom-up improvements, fractal innovation flows from a decisive top-down imperative. In this light, we can readily appreciate the tensions between the 2ASDU team and the assembly plant, which was shaped by its history with Japanese approaches and which witnessed the radical disruption of the improvement approaches set up in the project intervention mode.

8.2 Organizational Conditions of Fractal Innovation

We used the various episodes analyzed in the Kwid's development history to identify the three key ingredients that enable fractal innovation: concurrent engineering, intrusive management, and heavyweight project management.

8.2.1 Concurrent Engineering

Fractal innovation was not confined to just one of the various areas of expertise that contributed to the design of the new products but was located at their intersection. The Product/Process engineers of the 2ASDU project office, along with the buyers and managers who defined the requirements, could all legitimately lay claim to the Kwid's success, but none could take individual credit for its creation. In Part I, we saw how the "economy of the rupee" was based on the ability to explore potential cost-saving areas in a fully interactive and simultaneous manner downgrading functional requirements in relation to competitors in the specific target market, while questioning standards from the Product and Process engineering functions right through to market competitive strategy.

Behind this systematic exploration lay the ability to carefully measure the impact of each design variable on the overall cost reduction. Finally, there was also an ability to base decisions not on the name of a business theory—be it in marketing, engineering, or purchasing—but on the strength of an inter-business compromise over a specific project objective.

The project war-room,[9,10] a small space in which all the technical experts dedicated to the project worked together, was the first step in the physical organization

[9] Garel, Gilles. "L'entreprise sur un plateau : un exemple de gestion de projet concourante dans l'industrie automobile." *Gestion 2000*, no. 3 (1996): 111–134.

[10] Garel, Gilles. *Le management de projet.* La Découverte, Repères, Paris, 2011.

of the project based on this cross-functional strategy. The breakdown of the team into functional sub-assemblies and overall management through weekly team meetings was translated into the project organization chart.

However, we need to keep in mind that an entire automotive development project did not come down to a 15-member core project team. It mobilized hundreds of operators and heavy facilities (mock-up models, prototypes, test benches, etc.) in locations from India to France, Japan, South Korea, and now Brazil. Such a machine could not be governed by an "adhocracy,"[11] in which operators interacted when required. It must be remembered that in large automotive groups, functional divisions now simultaneously handle more than 10 projects varying in size and nature. In addition to the project team, which was 100% dedicated to the project and located in the project office, the activities of these various stakeholders had to be coordinated.

Until the 1980s, this coordination was carried out using standard development schedules, which organized the sequential intervention of various functions during project phases. Marketing set the target, Design provided its mock-ups, and then Product and Process Engineering transformed this design into a functional vehicle and production methods, while buyers organized the competition of suppliers based on technical specifications provided by the engineers.

The various design phases followed one another: the engineering-phase stakeholder waited for the pre-engineering–phase stakeholder to complete its work "in order to know what to work on." For example, Process Engineering only intervened to define the production process once Product Engineering had completed the design of the car, which would have started its activity only after the design freeze imposed by Design, etc. In fact, these sequential processes led both to sub-optimization (because the pre-engineering phase stakeholder lacked important information known only by the engineering phase stakeholder) and to numerous reworks (alterations) when the engineering-phase stakeholder discovered that the solution selected by the pre-engineering–phase stakeholder was not suitable.

However, although sequential planning was largely ineffective, it was still practical at the organizational level, because it forestalled protracted interactions between divisions under highly uncertain conditions—on the surface, at least. Of course, such an approach was incompatible with the search for a multi-variable compromise that characterizes fractal innovation.

A major transformation of the design processes arrived in the 1990s, when these sequential intervention schedules were transformed according to then-current engineering strategy (see Figure 8.1). This set up a much more continuous

[11] Mintzberg, Henry, and Alexandra McHugh. "Strategy Formation in an Adhocracy." *Administrative Science Quarterly* 30, no. 2 (1985): 160.

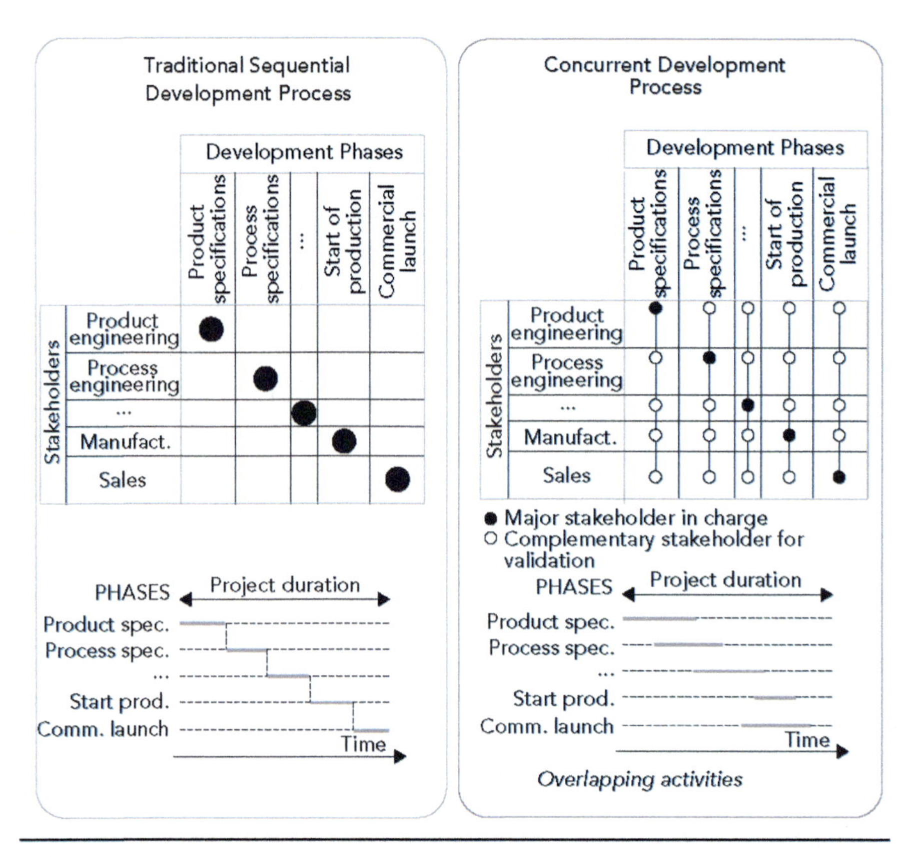

Figure 8.1 From sequential to concurrent business interventions. (Source: Midler own work.)

intervention of various experts during the progress of the project—to better anticipate the problems in the pre-engineering phase on one hand, and to prepare to receive innovations in the engineering phase on the other.

The other development of concurrent engineering, which the project led by 2ASDU had systematically implemented, was the development of the decision-making schedule, leading to the gradual "freeze" of various project variables.

One of the main difficulties of innovation projects is that they must contend with myriad uncertainties. Therefore, the art of managing the project was to link a learning process to these unknowns and a decision-making process that would ensure project convergence within a limited period of time. In the pre-engineering phase, anything was possible but little was known; in the engineering phase, a lot of knowledge was available, but the freedom to make the

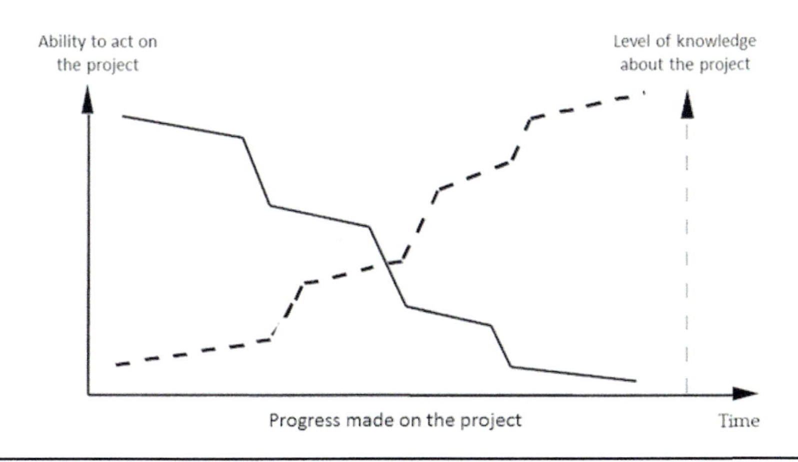

Figure 8.2 Project convergence as a dual learning and decision-making process.

most out of it had evaporated because of the irreversibility imposed by gradual "freezes." Figure 8.2 illustrates this dilemma.[12,13]

This diagrammatic representation shows how, with regard to the project, the speed of the project freeze is not the key to ensuring an efficient and rapid convergence. The freeze is, in fact, a decision made "while fully ignorant of the facts." Instead, efficient convergence is based on accelerated and early learning, while the delay in freezing is obviously not detrimental, within a certain limit (see Figure 8.3).

This concurrent engineering strategy, although clearly efficient when viewed from within the project, makes the activity of various technical stakeholders much more complex. Earlier, they were confined to their mission and specific local objectives, according to well-defined time periods, and operated on the basis of the frozen pre-engineering input data. Now they are immersed in numerous and diverse interactions, working on variables for which decisions has yet to be taken and on overall goals that relate well beyond their specific expertise.[14]

[12] Midler, Christophe. *L'auto qui n'existait pas. Management des projets et transformation de l'entreprise*. InterEditions, Paris, 1993. Nouvelle édition Dunod, Paris, 2012.

[13] Lundin, Rolf A., Niklas Arvidsson, Tim Brady, Eskil Ekstedt, Christophe Midler, and Jörg Sydow. *Managing and Working in Project Society: Institutional Challenges of Temporary Organizations*. Cambridge, UK: Cambridge University Press, 2015.

[14] Charue-Duboc, Florence, and Christophe Midler. "L'activité d'ingénierie et le modèle de projet concourant." *Sociologie du travail* 44, no. 3 (2002): 401–417.

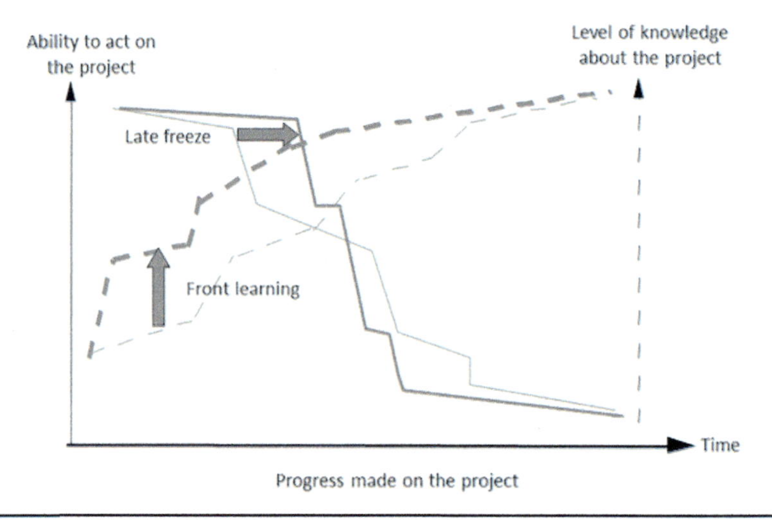

Figure 8.3 Concurrent engineering: early and accelerated learning, and delay in freezing the variables of the project.

8.2.2 *Intrusive Management*

The second catalyst for fractal innovation was the collective ability to learn. As mentioned earlier, the lack of learning activity in traditional automotive product development went hand in hand with the relaxed timescales. In the end, the standard solution would be selected just by playing for time, because there were no valid competitors. Hence the importance of being under constant pressure to explore new solutions and accelerate the learning loop: to quickly identify the potential for improvement; to mobilize dealer networks to conduct surveys; to shape the ideas received, overcome obstacles, and avoid traps; to quickly wind up debates that were going nowhere; and, finally, to know how to impose project decisions, even when they broke away from established practices and standards.

This ability to intervene in all the areas of the project was, as noted, embodied by the Program Director's large and deep automobile competencies as his management style.[15] However, it was deployed at all levels and across the entire scope, since it extended to the entire core 2ASDU team. This was based on several factors. First, team building: experienced and legitimate professionals

[15] Gérard Detourbet was elected "*Autocar Professional*'s Man of the Year 2015" by the Indian automotive professional community and "Man of the Year 2016" by the French Professional Award of the Automotive Profession, organized by *Le Journal de l'Automobile*.

with access to networks had been recruited. Second, a style of leadership was implemented that constantly reminded people of the need to go beyond the current situation. This leadership provided support as soon as an employee had a problem he or she could not solve alone; but it also did not hesitate to bluntly "straighten out" any stakeholders who, by their actions, were out of line with the philosophy of the project. This involved measures such as weekly *design-to-cost* meetings—the metronome measuring out this accelerated collective learning from beginning to end.

The product development professionals, of course, were already aware of this type of learning strategy. But where fractal innovation offered something original was evidenced in the systematic, almost obsessional dimension of its deployment over time (from the emergence of the design to its conclusion), and in its scope (encompassing design choices and requirements, plant locations, cable diameters, and the cost of screwdrivers on the assembly line, to reiterate some earlier examples). Such a sense of urgency is typical in engineering, once key milestones have been outlined. But it is not generally the case in the beginning of deployment phases. "Flagging up the crisis in advance" and revealing "the hidden urgency" of the exploratory phases were important management modes,[16] but they were also socially sensitive in their implementation. On the other hand, intrusive management forces the engineers to widen the scope of their attention to the entirety of the project. Usually, design energy quickly focuses on a handful of specific points: a new component that conferred a significant competitive edge; another that posed a quality risk. Less critical issues quickly slipped "under the radar." As the engineer in charge of the Electricity/Electronics field noted, "One of the benefits of working in India is that we count in rupees. In Europe, we would have been unaware of activities involving fractions of euros."

8.2.3 Heavyweight Project Management

It should come as no surprise that the practical implementation of concurrent engineering and micromanagement required strong policy management within the organization. Moreover, from the 1990s, management literature had stressed the importance of this "empowerment" of project role,[17,18] backed up by the

[16] Lenfle, Sylvain. "Exploration and Project Management." *International Journal of Project Management* 26, no. 5 (2008): 469–478.

[17] Clark, Kim B., and Takahiro Fujimoto. *Product Development Performance: Strategy, Organization, and Management in the World Auto Industry*. Boston, MA: Harvard Business School Press, 1991.

[18] Midler, Christophe. *L'auto qui n'existait pas. Management des projets et transformation de l'entreprise*. InterEditions, Paris, 1993. Nouvelle édition Dunod, Paris, 2012.

strong support of CEOs.[19,20] Twenty years later, negotiating the autonomy of the project was still a struggle, even though it was regarded as indispensable to mobilizing everyone's energies to the overall objective of the project. The XBA/I2 story showed that this battle was worthwhile, and the consequences of the 2ASDU team reporting directly to the CEO of the Alliance acted as a reminder of its efficiency. It is important to note that this happened in a company that had pioneered empowered project management, especially with the Twingo project.[21] This was proof of the extraordinary stability of the strategies set up in the permanent organization of firms when faced with the temporary organizations that made up individual projects.[22]

8.3 The Dynamics of Fractal Innovation and Design Processes: The Return of Creative Product Development

It would be amazing to discover that such a product development engineering practice was completely original. After all, wasn't design engineers' main role to optimize design variables in relation to a specific market target? What could be unique about a plan to develop a new "low-cost" product to target a gain of 50% against the best reference available at the time in the field? Wasn't that role already assigned to project management in the books cited earlier, which emphasized the "empowerment" of this function at the beginning of the 1990s? Could this all be a kind of organizational amnesia—a step back to what had been identified at the time as a great step forward in the creative ability of large companies? To answer these questions, we need to review the history of how car product-design processes were structured, especially at Renault.

In the 1950s and 1960s, the development of new products was the area in which innovation in automotive companies actually took place. Iconic products such as the Citroën DS were proof of this. From the beginning of the 1980s, the limits of this approach, in which innovations were tried out as part of product

[19] Burgelman, Robert A., and Leonard R. Sayles. *Inside Corporate Innovation: Strategy, Structure and Managerial Skills*. New York: Free Press, 1986.

[20] Burgelman, Robert A., Webb McKinney, and Philip E. Meza. *Becoming Hewlett Packard: Why Strategic Leadership Matters*. New York: Oxford University Press, 2017.

[21] Midler, Christophe. *L'auto qui n'existait pas. Management des projets et transformation de l'entreprise*. InterEditions, Paris, 1993. Nouvelle édition Dunod, Paris, 2012.

[22] Lundin, Rolf A., Niklas Arvidsson, Tim Brady, Eskil Ekstedt, Christophe Midler, and Jörg Sydow. *Managing and Working in Project Society: Institutional Challenges of Temporary Organizations*. Cambridge, United Kingdom: Cambridge University Press, 2015.

developments, were identified. New-product development periods were very long and became incompatible with marketing strategies that advocated more new products plus constant renewal of those already on the market. The introduction of experimental technical innovations put product quality at risk, and thus jeopardized brand image as well.

The following decades would see a gradual separation between exploration and the maturing of innovations and their application on the one hand, and new products under development on the other. Its dawn was the strong validation of product-development cycles by the person who would soon become the CEO of Renault, Bernard Hanon, at the beginning of the 1970s, and the creation of a research department in Renault in 1976. From then on, the collective creative activity of the firm would gradually become specialized and divided into three phases:

- The pre-engineering research phase, which was basically mobilized to obtain new generic technologies (such as robotics, materials, electronics, etc.) and turned more toward academia than to automotive engineering to do so.
- The advanced engineering phase,[23] during which the functional and/or technological "bricks" adding value to the vehicle (navigation system, driving support system, engine hybridization, an "intelligent seat," etc.) were explored and matured outside the product development cycle. This phase "ended" when these bricks were technically mature enough[24] to avoid putting the development of the vehicle at risk, and when it was clear that their customer appeal would enhance the competitiveness of the product.
- The vehicle engineering phase, which ensured that the product and process system was correctly developed. This was divided into a short pre-project phase that aligned the global variables, and then the more detailed development of the product and process.

In this context, development engineering's mission was to reduce the costs and lead times of developing the integrated system—the vehicle—while assuring product quality following the commercial launch of the car. The "industrialization" of engineering of the 1990s and 2000s was part of this mission: the maximum reuse of existing validated solutions; increasing modularization associated

[23] Beaume, Romain. *L'Ingénierie Avancée & les Programmes Pilotes ; Les Dynamiques d'Innovation Automobiles.* Thèse de l'Ecole Polytechnique, Paris, 2012.

[24] Maturity measured with the *Technology Readiness Level* (TRL) criteria. https://en.wikipedia.org/wiki/Technology_readiness_level

with outsourcing design to sub-assembly suppliers[25]; the use of CAD tools to automate development; etc. And, hence, there would be significant reduction of product-development lead time and costs during this period, as well as quality improvements at the start.

Clearly, innovation had hardly any place in this engineering, which was increasingly "rules-based design"[26] and focused on the problem—very difficult in itself—of integrating off-the-shelf solutions based on the most standardized operating modes possible. Innovation represented the additional costs of research, as well as risks related to quality and lead time. Yet at the same time, competition in mature, saturated markets was placing more emphasis on "visible" innovations that could differentiate products and lift them out of the "red oceans"[27] of price wars and trade discounts.

The mid-2000s saw the creativity issue return with a vengeance. At Renault—urged on by Yves Dubreil, the charismatic innovator with the Twingo to his credit—the Department of Innovation and Automotive Engineering (DISA) was created in 2005. Its task was to take charge of exploring breakthrough vehicle concepts within the Vehicle Engineering Department that Dubreil managed at the time. The aim was to restore the firm's creative ability in the complete vehicle space, which had been one of the key assets of the brand in the past, with breakthrough concepts such as the R16, the R5, the Espace, Twingo and the Scenic. Thus, DISA was behind the concept of Twizy, the small electric vehicle launched in 2011 as part of Renault's Zero Emission program.

DISA was a precursor to the Cooperative Innovation Laboratory (LCI), which incubated the Kwid in its emergence phase. At the same time, other initiatives were launched to reopen other areas that were conducive to creativity. The creation of a "fablab" within the Technocentre (the "Creative Lab"), an exhibition and communication area on innovation (the "Innovation Room"), and a community system ("Renault Creative People") helped to cultivate employees' ideas beyond their normal engineering activity—just as it had at Google. Furthermore, the Renault innovation community, managed by Dominique Levent, would connect internal innovators with the operators of other companies with similar issues, as well as academic experts from the innovation field.

---

[25] Fourcade, François, and Christophe Midler. "The Role of 1st Tier Suppliers in Automobile Product Modularisation: The Search for a Coherent Strategy." *International Journal of Automotive Technology and Management* 5, no. 2 (2005): 146.

[26] Le Masson, Pascal, Benoît Weil, and Armand Hatchuel. *Strategic Management of Innovation and Design*. New York: Cambridge University Press, 2012.

[27] Kim, W. Chan, and Renée Mauborgne. *Blue Ocean Strategy: How to Create Uncontested Market Space and Make Competition Irrelevant*. Boston, MA: Harvard Business Review Press, 2005.

These Renault initiatives were far from unique in the industrial world. The late 2000s and early 2010s saw more and more teams in charge of driving breakthrough innovation at large firms, including the innovation fields at Valeo,[28] Ilab at Air liquid,[29] and OpenLabs at PSA. These dynamics were also based on the mobilization of various academic communities, with the development of innovative and ambitious ideas such as the Concept Knowledge theory[30,31]; the deployment of design methodologies under the banner of Design Thinking,[32] promulgated by prestigious institutions such as Stanford; or the trends of organizational ambidexterity[33] and exploration projects[34,35] in the field of organizational theory.

Once again, it was noted that the creative effort was all "upstream": it all happened in the survey and distant exploration phases, at the concept stage, and during the incubation of "out-of-the-box" concepts. There was still a great deal of potential to creatively transform the design processes of large companies. But at the same time, limits were becoming apparent. Creative ideas incubated in advance struggled to pass through regulated development and reach the market under competitive conditions. Between the mock-ups or sparkling "proofs of concept" and the products sold, the development phase appeared to be an insurmountable barrier. When it left the LCI, the XBA project was an appealing design that had to demonstrate its economic feasibility once more by delivering a 30% gain against its first "bottom-up" production-cost figure.

---

[28] Ben Mahmoud-Jouini, Sihem, Florence Charue-Duboc, and François Fourcade. "Favoriser l'innovation radicale dans une entreprise multidivisionnelle : Extension du modèle ambidextre à partir de l'analyse d'un cas." *Finance Contrôle Stratégie* 10, no. 3 (2007): 5–41.

[29] Ben Mahmoud-Jouini, Sihem. "Innovation Units within Established Firms. Towards a Cartography." *IPDMC Proceedings,* Copenhagen, June, 2015.

[30] Hatchuel, Armand, and Benoît Weil. "C-K Design Theory: An Advanced Formulation." *Research in Engineering Design* 19, no. 4 (2008): 181–192.

[31] Le Masson, Pascal, Benoît Weil, and Armand Hatchuel. *Strategic Management of Innovation and Design.* New York: Cambridge University Press, 2012.

[32] Brown, Tim, and Barry Katz. *Change by Design: How Design Thinking Transforms Organizations and Inspires Innovation.* New York: Harper Business, 2011.

[33] Tushman, Michael L., and Charles A. O'Reilly. "Ambidextrous Organizations: Managing Evolutionary and Revolutionary Change." *California Management Review* 38, no. 4 (1996): 8–29.

[34] Lenfle, Sylvain. "Exploration and Project Management." *International Journal of Project Management* 26, no. 5 (2008): 469–478.

[35] Midler, Christophe, Catherine P. Killen, and Alexander Kock. "Project and Innovation Management: Bridging Contemporary Trends in Theory and Practice." *Project Management Journal* 47, no. 2 (2016): 3–7.

There was still a mountain to climb: XBA contained no technological miracle that would carry it through to launch. It was only through fractal innovation during the development stage that this challenge was met. With a normal regulated development approach, the project would have been quashed at the outset, faced with a challenge that nobody would have dared to take up. And if, by exception, the project passed the "green light" to development phase, it would end up mired in extra costs resulting from the accumulation of standard solutions and constraints.

For a decade, the spotlight of innovation remained fixed on the pre-engineering design phases, as if collective creativity could spring from nowhere else. This was true of both academia and industry. The best that could happen in the rest of the story—the development project—is that the creative achievement obtained at this initial stage would be preserved. In contrast, the 2ASDU team's implementation approach offered to give product development back to the creative dimension that it had lost because of the previous strategy of streamlining design. This approach could be used to open up new horizons and bring authentically original products to market.[36] This was possible because the ingenuity mobilized to cut costs could, in another project, be used in other performance dimensions. At Toyota, the first Prius project was another example of a development that had implemented fractal innovation to reduce the fuel consumption of the new hybrid technology.[37] In our research on the electric vehicle, we showed how decisive the engineering phases of the project were in expanding the sphere of innovation to the more global electric mobility system, including the car, infrastructure, and services required for this new technology.[38]

Even beyond the development of the innovation that introduced the breakthrough, we have shown in several examples from the automotive industry and elsewhere that the competitiveness of innovation management was associated with efficient project lineage management,[39] which would deploy[40] the initial

---

[36] Gemünden, Hans Georg. "When Less Is More, and When Less Is Less." *Project Management Journal* 46, no. 3 (2015): 3–9.

[37] Itazaki, Hideshi. *The Prius that Shook the World: How Toyota Developed the World's First Mass-Production Hybrid Vehicle.* Tokyo: Nikkan Kogyo Shimbun, 1999.

[38] Pechmann, Felix von, Christophe Midler, Rémi Maniak, and Florence Charue-Duboc. "Managing Systemic and Disruptive Innovation: Lessons from the Renault Zero Emission Initiative." *Industrial and Corporate Change* 24, no. 3 (2015): 677–695.

[39] Maniak, Rémi, and Christophe Midler. "Multiproject Lineage Management: Bridging Project Management and Design-Based Innovation Strategy." *International Journal of Project Management* 32, no. 7 (2014): 1146–1156.

[40] Ben Mahmoud-Jouini, Sihem, Thierry Burger-Helchen, Florence Charue-Duboc, and Yves L. Doz. "Global Organization of Innovation Process." *Management International* 19, no. 4 (2015): 112–120.

concept across an extended product range and diverse markets.[41] This approach could preserve the original DNA of the initial concept from project to project over time, while adapting and developing it based on new learnings and contexts. Even before the Logan was launched in 2004, and up to its renewal in 2011, the engineering teams of the Entry program made a series of different products, from the Sandero to the Duster, increasingly profitable by taking advantage of short-loop market feedback.[42] Conversely, standard-platform strategies compelled them to anticipate all the potential marketing scenarios at the start, and this led to the inclusion of constraints that would, unfortunately, often be perceived *a posteriori* as being inadequate. Product-lineage strategies, shown to be profitable by the Entry family, could not be separated from creative development engineering. Because of the scope of the research, we could not study this line management on the Kwid as we could on the Logan. The launch of the Redi-GO, however, like the current deployment of the Kwid in Brazil, showed that such an approach was still being used.

In this respect, if the lessons are learned, the Kwid could be much more than an anecdote about the ingenuity of automotive pioneers. It could herald a new stage in the large-scale organization of collective creativity.

[41] Ben Mahmoud-Jouini, Sihem, Florence Charue-Duboc, and Christophe Midler. *Management de l'innovation et globalisation, enjeux et pratiques contemporains*. Paris: Dunod, 2015.

[42] Midler, Christophe. "Implementing a Low-End Disruption Strategy Through Multiproject Lineage Management: The Logan Case." *Project Management Journal* 44, no. 5 (2013): 24–35.

Chapter Nine

The Maturation of a "Trickle-Up" Approach of the Automobile at Renault

The success story of the Kwid seemed to confirm the unique strategy of the Alliance and its members. In today's automobile industry, it is accepted practice that innovations are integrated in high-end ranges first, before being extended to the rest once they become cheaper and better controlled: a "trickle-down"[1] trajectory. The brand image and, indeed, profits are also believed to behave accordingly. With the Logan and the Kwid, innovation aimed to maximize the odds of converting non-customers into customers by pursuing a "bare-minimum" approach, which would ultimately lead to upscaling the range. Thus, there was a focus on a "trickle-up" approach, in which the fundamentals were built based on customers with the greatest constraints. A major part of the engineering effort was concentrated on this point, and it was only

[1] In the American macroeconomic debate, "Obamanomics" is often characterized in the press as an attempt to create "trickle-up economics," in contrast to the trickle-down economics that have dominated since the time of Reagan. We reinterpretate this analysis from the specific perspective of the automobile sector in Jullien, Bernard, and Tomassso Pardi, "In the Name of Consumer: The Social Construction of Innovation in the European Automobile Industry and Its Political Consequences." *ERIEP* 3, 2013. http://revel.unice.fr/eriep/index.html?id=3338

by succeeding at this that profitability could be achieved. Kwid's story may indeed be the initiative that brings this approach to its phase of maturity, within Renault at least. In fact, while not neglecting its conventional ranges, which could be termed "mainstream," Renault has grown in strength since it launched the Logan in 2004 as part of its "Entry and sub-Entry" range. This new offering stood out both as a valuable addition to Renault's range in mature European markets and as a powerful tool in capturing high-growth emerging markets.

In its European markets in 2016, Dacia has generated a little over a quarter of Renault's total sales. In other markets in which Dacia and Renault coexist, such as the Turkish, Algerian, Romanian, or Moroccan markets, the situation is often reversed. In Morocco, for example, Dacia has 24.3% market share and Renault only 8%. Last but not least, in emerging markets in which Renault is present to a significant degree (Brazil, Russia, India, Argentina), it is the largest firm in the Entry range (and sub-Entry in India) and is in a strong, dominant position, if not a monopoly. It was thanks to the Logan that Renault truly became an intercontinental firm, thus achieving an objective that had previously been more of an aspiration than a reality. Hence, in 2016, as had been the case for several years, the sales of Renault's models surpassed those of Dacia.

Since 2004, Renault's innovative approach to products and markets, along with its adoption of the design methods described in the previous chapter, has thus become pivotal in terms of volume, growth, profitability, and the capability to establish itself as a major player in emerging markets. It has become clear within Renault that the Entry range, and the methods used to assure its success, are major assets. The fact that Nissan has put its brand on models from the Logan line in certain markets since 2007, and that it also chose to do so in India and Russia for the Duster (branded the Nissan Terrano), confirms that the value of this range is worth leveraging within the Alliance. The launch of the Datsun embodies this ambition, and the fact that the new CMF-A platform has been developed to be used as the base for a Renault model (Kwid) and a Datsun model (Redi-GO) make it a reality. The bottom-up approach has thus established itself as a key strategic tool to develop and penetrate emerging markets, and it is enthusiastically supported by Carlos Ghosn, CEO of the Alliance.

Throughout the same time period, for all the Alliance brands, in particular for Nissan, Infiniti, Samsung, and Renault, the more mainstream products nonetheless remained of vital importance. The intention to launch them in large mature markets, as well as in certain accessible emerging markets such as China, continued to dominate the strategic discourse of the group. This is what bolstered the key investments and major business rules Renault and Nissan provided to each other—first separately and then, gradually, together. It is in this environment that the programs and teams managing the Entry range continued to promote—both within Renault and even the wider Alliance—a form of counterculture or alternative to the dominant narrative. Despite the volumes and profits that the Entry range was generating, it thus remained challenging to integrate it completely.

As the Kwid story emerges from discussions ahead of the decision to move ahead to its commercial launch, it is striking that many of the questions raised at every step had already been asked for the Logan. The most iconic illustration of this is the fact that the Logan project only originated after earlier studies into designing with conventional methods of "decontenting" an established old product were unsuccessful. Recognizing that the low-cost target could not be achieved by using such a method, it was only then concluded that it would be necessary to create a unique project focused on this target, in order to open up the scope enough to include the architectural choices required to achieve it. The exact same discussions occurred again, in Nissan as well as in Renault, when plans were being made to develop an Entry range for all emerging markets (and not another Entry for mature markets). All the same hypotheses based on "recycling" older platforms were thus examined by both Datsun and Renault before the same conclusion was reached: given the cost constraints imposed, only a single dedicated program would allow the development of a convincing vehicle.

The Kwid and CMF-A were built without support from corporate engineering departments, top management, or operational departments of the parent companies. Support from head office was, and continued to be, a problem. Between 1998 and 2004, unwavering support from Louis Schweitzer was accorded to the teams working on the Logan project, and this went a long way to winning over the skeptics. The same process was necessary for Renault and the Alliance between 2011 and 2016: Carlos Ghosn handpicked Gérard Detourbet and gave him the freedom and space required to work with or without support from corporate functions. Ghosn reiterated whenever he had to that he was fully behind the project and that he had complete faith in the Program Director to deliver.

This single-mindedness, orchestrated and supported by the CEO, demonstrates the difficulty faced by large organizations, such as car manufacturers, to integrate Entry ranges and the associated radical design-to-cost decisions that involve countless transgressions of the rules in place within both Renault and Nissan. In line with the theme of the ambidextrous organization, extensively discussed in management literature, we could say that Renault's two successive leaders afforded it a form of "structural" ambidexterity[2,3,4]: On the one hand, the CEO takes a partly oppositional stance by supporting a structure that defends the innovative project. On the other, however, they continue to support the core of the organization, which exists to pursue enhanced efficiency in conventional manufacturing.

Nevertheless, as Carlos Ghosn himself often insisted (and even mentions in his preface), such projects, even if they are based in sound strategy, are pointless if they are not well executed. For the Kwid to see the light of day—as every chapter of Part I indicated—the project had to mobilize not only all its own resources (including those that management assigned outside 2ASDU), but also a good number of corporate resources around the fundamental objectives of the project (the four famous USPs). It has been said of the Twingo that, "Innovation is the product of disruption, that is obvious; but it is also the product of memory."[5] The ambidexterity of Renault is more than just the consequence of an innovative group within the top management of a conservative form. It is also the consequence of a specific combination of proprietary resources that have acquired cumulatively through experience throughout a career.

The nature of fractal innovation, which made the project possible, implied that the internal and external elements of the program themselves were ambidextrous. To succeed, the people who were mobilized to achieve the highly specific objectives of the program had to draw on everything they knew in order to disrupt conventional thinking and practice. However, they also had to do so without exposing the company to too many risks or wasting too much time in the program.

As a result, integrating the Entry range overall, and in particular the Kwid, within Renault was complicated. On the one hand, it needed autonomy and support from the CEO so that the projects were successful. On the other hand, it was still in fact Renault, with all its resources and accumulated expertise, that were mobilized through these projects. This conflict structured the relationship between the Entry range and the mainstream ranges—and, ultimately, the strategic "position" of the Entry range in the trajectory of the company. There

---

2 Tushman, Michael L., and Charles A. O'Reilly. "Ambidextrous Organizations: Managing Evolutionary and Revolutionary Change." *California Management Review* 38, no. 4 (1996): 8–29.

3 Benner, Mary J., and Michael L. Tushman. "Exploitation, Exploration, and Process Management: The Productivity Dilemma Revisited." *The Academy of Management Review* 28, no. 2 (2003): 238–256.

4 Gilbert, Clark G. "Change in the Presence of Residual Fit: Can Competing Frames Coexist?" *Organization Science* 17, no. 1 (2005): 150–167.

5 Midler, Christophe. *L'auto qui n'existait pas. Management des projets et transformation de l'entreprise.* InterEditions, Paris, 1993. Nouvelle édition Dunod, Paris, 2012.

were both conflicts and unity, and the CEO had to balance this ambivalence. Functions and even individuals had to switch from one philosophy to another, from one period to the next, and even, sometimes, from one moment to the next. This situation is described in the literature as "contextual" ambidexterity.[6,7]

Recent events in the history of product policy and organization decisions taken by Renault help to clarify this complex relationship. The disruptive nature of this story, the prevarications and constant back and forth, actually appear to have shaped a memory and an aptitude to overcome challenges whose characteristics were reflected in the Logan and then in the Kwid. Understanding the success of the Entry range requires understanding how its relationships with the company evolved and how these programs became integrated with the rest of the firm to become part of mainstream business-as-usual practices. This was possible in part because of the quality of the team Gérard Detourbet had assembled in Chennai, and in part because of his ability to mobilize the expertise of every function that reported to him. What emerges is a paradoxical institutionalization of ambidexterity and of the forms of coexistence between the mainstream and the Entry ranges.

9.1 Renault: Built to be Ambivalent

Prior research has tried to plot Renault's trajectory since the end of the Second World War, and to classify it in comparative categories to understand the dominant strategic rationales. Generally, such works highlight the instability of the firm's approach and the prevarications associated with it. While Renault's strategic choices may not initially have appeared as clear as those of VW or Toyota, the Kwid's story reveals that lack of strategic certainty can, in fact, help promote ambidexterity.

Michel Freyssenet,[8] for example, argues that between the 1950s and the 1960s, Renault switched from a "Taylorist" to a "Sloanian" model and sought to manage its exit from a quasi-mono model that emerged after the war to

[6] Gibson, Christina B., and Julian Birkinshaw. "The Antecedents, Consequences, and Mediating Role of Organizational Ambidexterity." *Academy of Management Journal* 47, no. 2 (2004): 209–226.

[7] Hargadon, Andrew B., and Angelo Fanelli. "Action and Possibility: Reconciling Dual Perspectives of Knowledge in Organizations." *Organization Science* 13, no. 3 (2002): 290–302.

[8] Freyssenet, Michel. "Renault, from Diversified Mass Production to Innovative Flexible Production." In *One Best Way? The Trajectories and Industrial Models of World Automobile Producers.* Michel Freyssenet, Andrew Mair, Koichi Shimizu, and Guiseppe Volpato (eds.). Oxford and New York: Oxford University Press, 1998, pp. 365–394.

undertake the progressive development of a range.[9] Failing to increase profits, Renault insisted on "pursuing a policy based on volume and expanding the range" for more than 20 years. At one point, the company managed to become the top carmaker in Europe, but fell so deeply into debt that in the mid-1980s it had to "lower its break-even point" and capitalize on profits by focusing more on the high-end segment than on volume.

This was the period during which the pursuit of quality became the leitmotif, under CEO R. H. Levy's leadership. It was also when Renault attempted to partner with Volvo. Between the recession of the mid-1980s and the fall in demand in Europe in 1993, Renault went through a period of success centered on a range of models that were practically all mainstream, apart from the Twingo, which was launched in 1993. The R19, for example, was launched in 1988, the Clio in 1990, the Safrane in 1992, and the Laguna in 1993. Renault's efforts to meet the quality imperative in a context of strong European demand finally paid off. It appeared to have found its strategic direction.

However, after the failure of the merger with Volvo and the sharp fall in European demand in 1993, history seemed to repeat itself. When Louis Schweitzer became Chairman and CEO in 1993, he insisted that Renault could make a difference in the context of saturated demand only if innovation was centered around models like the Espace and the Twingo. To sum up the 1990s, Michel Fresseynet wrote:

> *After Volvo's exit and the economic downturn, Renault shifted its product policy to commercially innovative products with individual personalities. It became profitable again for the following nine years with a unique range of products whose quality had been significantly improved. Renault had reached a point where it was no longer in debt and transformed itself into a private enterprise.*

From Freyssenet's point of view, what this meant was that Renault's Sloanian endeavors had never been very promising and that, *a posteriori*, the difficulty had clearly only been overcome by allowing innovation in the area of design. This generated models with distinct characteristics that created new segments and, for a while at least, assured the company a strong, dominant position. Freyssenet explains that Louis Schweitzer designed the Mégane range to innovate on at least at two levels:

- The range of bodies would be particularly broad because, unlike what had been done for R19 and R21, six variants were proposed (five-door, saloon, estate, convertible, coupé, passenger van).

[9] Loubet, Jean-Louis. *Citroën, Peugeot, Renault et les autres. Soixante ans de stratégies.* Paris: Le Monde-Éditions, 1995.

- For the C segment, the Scenic passenger van sought to revive the concept of the Espace and corresponded to the creation of a new segment.

In fact, the Scenic became a commercial success that far exceeded expectations and became a pillar of Renault's success for several years. Launched in October 1996, the compact passenger van accounted for 44% of Méganes produced in 1997. In 2000, Renault produced 404,000 Scenics and 392,000 Méganes. Scenic 2, the subsequent version launched in 2003, mirrored its success: 390,000 were assembled in 2004.

It could be concluded that Renault had hit upon a winning strategy—differentiation through design—and that it could thus eliminate the inherent difficulties of the Sloanian strategy adopted in Europe by VW and PSA. However, the success of the Scenic must be weighed against Renault's failures in its high-end range during the same period. These include the Avantime and the Vel Satis, as well as in the B segment, where the Modus failed to create a niche. It is also argued by Schweitzer that the Scenic's success stemmed directly from a strategy of sharing standardized components, which allows the targeting of a niche at no additional cost and/or a shot at innovation without using disproportionate means.[10] From this perspective, Schweitzer emphasized that, despite ridiculously low sales levels of below 22,000 in its best year, the Vel Satis incurred considerably smaller losses than the Safrane, which actually sold three times as many units. Even though both cars shared the same platform as the Laguna and the Espace, a customized platform had to be developed for the Safrane.

Louis Schweitzer[11] points out that in the year before Carlos Ghosn took over, Renault was on a par with VW in Europe. Its model was more in line with the Sloanian/VW approach and was set up to maintain a large range and renew it as frequently as possible, from segment A to segment E. At the end of the 1990s, Schweitzer gave priority to bottom-up rather than top-down development in his pursuit of volumes. He believed that in order to compete with VW, it would be better to outperform Skoda than to try to challenge Audi, thus sharing the convictions of one of his great predecessors, Pierre Dreyfus, who launched the R4 in 1959.[12]

In terms of "strategic intent" and, subsequently, throughout the organization of design and production, the desire to innovate had never been so strong. This meant marginalizing the mainstream range and/or making trade-offs in

[10] Schweitzer, Louis. *Mes années Renault: Entre Billancourt et le Marché Mondial*. Paris: Gallimard, 2007.

[11] *Ibid.*

[12] Fridenson, Patrick. "Le procès de la R4 n'aura pas lieu." *Entreprises et histoire* 1, no. 78, (2015): 147–149.

line with what Boyer and Freyssenet[13] referred to as an "innovative and flexible" strategy, exemplified by Honda. The aim was to build success and profits based on just a few products or unique technologies, which would be designed to differentiate the brand and reap the "benefit of innovation." It required substantial flexibility since, in order to be successful, Renault had to be able to increase the volumes of its products very quickly when the market responded favorably—or, alternatively, quickly revise its strategies when there was a drop in demand without sustaining heavy losses. Similarly, it made sense to be somewhat integrated with suppliers, to share the risks. Finally, it was also preferable not to customize assembly sites too much around a single model.

In reality, Renault only did all of this to a very limited degree. Louis Schweitzer cited innovation in the list of strategic priorities, but his main focus remained volume produced and market share acquired. Similarly, on his arrival, Carlos Ghosn prioritized profitability and high margin sales to strategic sales. In every case, innovation was a strategic focus only when it emerged, or *de facto*. Functions were primarily structured and organized around the core of the range, which was often inevitably mainstream, since it targeted volume and/or profits. From management's point of view, innovation was a risk you could take because you knew how to manage "business as usual" properly. Profits, when they came, were primarily levers of development for business as usual, and only then additional means for taking new risks when innovating.

This situation led to conflict between innovative project teams and the teams supporting them on one side, and the core team and/or reorganization project teams on the other. *L'auto qui n'existait pas*[14] highlights how the Twingo project, which emerged from a laboratory for concurrent engineering, was confronted with this longstanding conflict even though it inherently proposed a way to overcome it at the same time by managing through projects. Project management organization allows access to resources from well-established functions that need to be used and justified for mainstream initiatives, as well as for more transgressive projects such as the Twingo and the Logan, which require an interdepartmental approach that traditional organizations find very difficult to manage. This form of organization was becoming almost universally regarded as the key to responding to the core requirements of new competitive spheres, both inside the automotive industry and beyond it. It became central at Renault because it integrated the heightened ambivalence in relation to product strategy and presented a non-explicit yet regulated way of organizing it.

[13] Boyer, Robert, and Michel Freyssenet. *The Productive Models: The Conditions of Profitability*. London: Palgrave, 2002.

[14] Midler, Christophe. *L'auto qui n'existait pas. Management des projets et transformation de l'entreprise*. InterEditions, Paris, 1993. Nouvelle édition Dunod, Paris, 2012.

9.2 From the Logan to the Kwid: A Marginality Less and Less Marginal

The Logan began as a marginal project with limited ambitions. As Louis Schweitzer explains: "We had appointed a Project Director [Jean-Marie Hurtiger] who was not a star at the time. He really proved himself in this job, but he could not be a star, because the project did not entice stars." The others assigned to the project were highly experienced, but their career paths had had many ups and downs. Many among them had spent a lot of time abroad, often outside Europe, at a time when Renault was barely internationalized but had announced big ambitions for its "intercontinental" projects. In reality, they were confined to the quest for additional volumes for products designed for Europe, where the volumes sold had already generated profitability.

Such individuals were used to fighting against Corporate, which wanted them to apply standards and recommended practices that they felt were not well adapted to their environments. Following the acquisition of Dacia and given the potential of the Logan, they finally had a real opportunity, in addition to the CEO's support. Their age and seniority, which could possibly have been considered an impediment, was perceived as a sign of experience and an assurance that they would remember all the methods and tools used in the company 30 years ago, as these were considered to be the ones they would have to use in Romania. The plant, in particular, would have to assure a level of quality that would normally require a whole range of automated equipment that was not available to them. The Logan would represent a return to the "bare minimum" strategy and operate under rules or references that had disappeared during the years of "sustaining innovations" that had been integrated into even the cheapest of Renault's products.

However, it was not simply a case of handing over the reins to veterans, with a brief to revive former methods and tools. Right across engineering, production, procurement, and sales and marketing, the teams needed to demonstrate a high level of creativity.[15] Many young executives were also hired, and they quickly understood that this project offered the potential to acquire and apply new skills faster than in other, more standard projects. As the project expanded, top management and project managers realized that they were going to have to increase the scope of both the industrialization plan and then the number of models. The Logan project thus began moving from a status of "project" to that of a "program." Executives that were already involved communicated this information to convince prospective candidates to join.

[15] Jullien, Bernard, Yannik Lung, and Christophe Midler. *The Logan Epic: New Trajectories for Innovation*. Paris: Dunod, 2013.

Across the areas of design, purchasing, sales and marketing, and manufacturing, executives who chose to join the Logan adventure look back on that time developing the Entry range as a period of rejuvenating challenge that, more than anything else, allowed them to show what they were capable of. Their experience reflected the "intrapreneurial" nature of the system—a feature of ambidexterity—that was designed initially by the "Romanian story" and ended up strengthened by its success. However, it also embodied an "Entry method," structured around meta-rules[16] discovered and refined along the way: where marginality, autonomy, and the clarity of the ultimate goal (a truly modern car at €5,000) offered just enough direction to define the project and channel the energies around it. The generic principles, structured around design-to-cost and adaptability to the local industrial and commercial contexts, stabilized and shaped the different functions.

These initial conditions allowed the workforce to grow and new employees to find their feet quickly. They also made the program intelligible and, to some extent, predictable for the different functions involved. Gérard Detourbet, backed by the priorities that he had asked management to validate, would challenge any rule that did not fit into his roadmap and could not be supported logically. If he was invited to a meeting where such rules were to be introduced or discussed, he responded systematically: "For all the other programs, if you want. But not for mine."

This is what managers became used to experiencing between 2000 and 2011, and particularly from 2006 to 2007 onwards, when sales of the Entry range exceeded 250,000 vehicles per year, three major plants were switched to manufacture the Logan, and the Sandero was announced. As opportunities increased, less was heard from those wishing to promote the status quo. The need for mediation dwindled, and the "double standards" that some found difficult to tolerate at the beginning became accepted practice. Likewise, fears that Entry veterans would become as unruly as their managers waned as they smoothly returned to the fold. Coexistence resumed.

While the mainstream operational managers did not necessarily accept such "defeats" gracefully, nor did they see them as fully justified, they did realize that they no longer needed to fight certain battles which they were bound to lose. The program had thus secured the necessary autonomy that would be further cemented by its results in terms of volumes and profits. The key lay in cooperating enough to ensure that the necessary resources were assigned to the program and allowing its unique qualities to attract a sufficient number of technical experts.

[16] Jolivet, Françoise, and Christian Navarre. "Grand projets, auto-organisation, méta-règles: vers de nouvelles formes de management des grands projets." *Gestion 2000* 19 (1993): 191–200.

Such was the context within which the Kwid took shape in 2011. Supporters of the Entry range had multiplied. Gérard Detourbet was no longer criticized, as he had been at the time he had accepted Louis Schweitzer's suggestion to join the Romanian team, and many countries or regions had already endorsed the program. Any conflict remaining was no longer between a marginal project such as the Logan and the rest of the company, but rather became a cross-functional divide between those who had experienced the program and gained from it and those who had not. Admittedly, the integration was incomplete, but it was well underway, and the organization of activities by programs provided a useful framework for decision making. The program tended to "freeze" the design early in order to allow more time for negotiation with suppliers; and, at the same time, designers downstream were heavily involved in the modifications required by the design-to-cost approach. Other programs considered this form of organization unnecessary. By this time, it was no longer Gérard Detourbet who managed the Entry program, but Arnaud Deboeuf, who was very close to Detourbet but had a more accommodating personality.

Consequently, the company appeared to be "ready for the Kwid," although it was really only halfway there. What was still required was far more substantive and comprised at least two levels of complexity. First of all, it involved an Alliance project—the first real one of its kind—and its leader was, from that point on, more than simply a program manager. He was "outside the organizational chart" and reported directly to Carlos Ghosn as the leader of the Alliance. In addition, it was announced that Gérard Detourbet was the Project Leader, and as he was past retirement age, he was not going to return to Renault. This unnerved functional managers, who feared the consequences of Detourbet's independence from organizational constraints. Secondly, the project was not confined to the development of a vehicle but involved the development of an entire platform, with two engines and two gearboxes that constituted the basis for the two models to begin with (for Renault and Datsun), and would undoubtedly be followed by other models. This meant that the organization transferred to Chennai in India not only a huge number of project hours but also some of its top engineers and a significant part of Renault's future product range.

This led to resistance and subsequently to the difficulties in recruiting and "completing the organization chart" in 2012, described in Part I. Similar problems arose when it was necessary to ask Corporate functions for studies on specific issues or, as was the case for acoustics, call on project managers to help deal with setbacks that arose because decisions had been taken in a context that was by definition riskier than usual. On the whole, however, internal relations were less tense than on the Logan project. The managers of the different areas of the project remained on good terms relative to their respective functions, probably as their seniority allowed them to rely on the quality of the interpersonal relationships they had developed over the years and the projects on which they had previously worked. Another reason was that most functions contributed to the Kwid as much as they learnt from it, and, even though the project was shipped off and developed far away, it remained a Renault project.

9.3 A Crew of Real Renault Fighters

In describing the project team that existed between the production of the first test Kwids and the launch of the car, one team member suggested that if the Technocentre in Guyancourt was an ocean liner, the team would be a fighter plane. Among the Design team and LCI managers, everyone was "on top of their game." The "fractal" nature of the work meant driving forward an ambitious design-to-cost strategy in a short time and on a large scale. Its documentation had to be handled in an inventive and hands-on manner, both for different methods and for all levels.

One of the special features of the project was that it involved the development not only of the new CMF-A platform, but also of a new engine that was soon to be followed by a more powerful new version and of a new gearbox for which a new robotized version was quickly to be proposed. Conventional wisdom would not compound all these difficulties at the same time as the new process of innovation management described above was being adopted. For example, when Renault developed a significantly innovative diesel engine, the R9M, widely known as Energy dCi 130, it was first fitted in the Scenic, even though that vehicle had already been in production for six years.

To add to the nature of the challenge, it was being taken on in an Indian context, in which the teams were mainly made up of engineers who had never developed vehicles or "complete" powertrains, and in which suppliers had never been asked to face similar questions. The team clearly needed people who were among the best in their field. Having spent forty years at Renault, Gérard Detourbet knew both who these bright sparks were and which of them would rise to the challenge. He had already worked with many of them on the Entry program or elsewhere. He trusted them and was confident they could train their teams. Kwid's Chief Engineer until its launch in 2015, Jean-François Vial, as well as the Engine and Powertrain Managers, were old acquaintances of Detourbet, with whom they had worked when he was the Powertrain Leader, and even prior to that. One of them had spent his entire career at Renault, having been recruited as a technician and promoted to management, thus affording him significant legitimacy in the eyes of both his fellow engineers and the technicians on their teams.

These team members had not moved from one highly innovative project to another, nor had they arrived from one expatriate posting to another. They had spent most of their careers in Guyancourt, at the heart of their respective functions and in mainstream projects. The norms they had to challenge in order to complete the highly specialized specifications for the XBA/I2 project were rules they knew thoroughly. Not only had they applied them over the years, they had often participated actively in designing them; and while they had found them restrictive at times, they knew why they were necessary.

For these reasons, it would not be correct to describe the evolving Kwid project as a start-up that was somehow marginal to Renault's mainstream organization. What Renault achieved, no start-up could ever have hoped to achieve. The whole project, and the hands-on management that made it possible, was an outcome of Renault expertise. That expertise was personified by those Renault employees who were sent to India to leverage their competencies, including that of how to "challenge Corporate" when it was necessary.

The Design Department—omnipresent in the project—embodied this dynamic, and this proved the key to the success of Gérard Detourbet and Arnaud Deboeuf's adventure. Because the quest to cut costs meant leaving a wide range of technical options open over a fairly long period, the integrity of interior and exterior design choices might have been affected by the choice of solutions that was finally retained by Renault or its providers to satisfy the very demanding cost target. The 2ASDU team focused on this "downstream" issue in close cooperation with engineering and the program leader. While each decision was dominated by an obsessive focus on every rupee spent, calculated in detail—and often with fury—every Friday by the Project Manager, the team also maintained extremely strong links with Corporate, which also validated or invalidated its work on a weekly basis.

For Design, as for most other functions, the nature of the project and the speed of its execution were completely new. Nonetheless, it still drew on one of its members' core competencies, which was the ability to interact with other functions and to build a constructive dialogue based on an underlying economic rationale. This dual competency is largely related to project management and what is called the "product-process combination." It was particularly practical for small cars—not only the Kwid—if a group such as Daimler (Mercedes) partnered with Renault to develop its new Smart or if the Alliance mobilized people from Renault and Nissan for A and B.

Unlike the Program Manager, the members of the fighter team knew that they were going to return to Guyancourt and were anxious about what would happen when the time came to reboard the "ocean liner." During their Kwid years, therefore, they wanted not only to satisfy the needs of the program and its leader, but also to avoid burning their bridges with those who had been their colleagues before, and would be again. Even as they worked against the rules, they knew they would have to submit to them once more, and the experience challenged them to find a way to accommodate this ambiguity.

They thus learned how to bend the rules by drawing on their extensive experience of those rules and of their in-depth understanding of why the rules had been made in the way they were. It was their experience that allowed these managers of different functions to challenge the rules, to trace their origins and weigh the implications of adapting them, and, above all, to pinpoint which other rules would be impacted by such changes. Cabling was the classic example of this, in that it facilitated a controlled form of rule-breaking that still allowed the team to land on their feet. Thus, just as the image of a start-up is at odds with reality, the metaphor of a blank slate also rings only partly true. "Blue sky" innovative explorations could not have resulted in a technically, economically, and commercially sound vehicle so quickly; team members leveraged their proven experiences and rules, even if only to move beyond them.

As has been shown in earlier criticisms of Taylorism, when the work of manual laborers is tightly supervised, they need to mobilize significant creativity in order to be productive.[17,18] In the case of the Kwid, although it may appear that the team was engaged primarily in exploitation, it is important not to forget that design activities are, by definition, always exploratory to a degree.[19] This was especially true in a company such as Renault, which struggled to stabilize its options and often explored solutions that it did not ultimately implement, or implemented only temporarily. It appears that these apparently fruitless phases of exploration nonetheless remained in the memory of the organization and its workforce and were available to be mobilized for a project like the Kwid. As a result, in addition to the interdepartmental cooperation and the low-cost culture highlighted earlier, the importance of the ambidexterity exhibited by technical experts needs to be recognized as a key factor in the success of the project. This was cultivated from within a corporate memory bank that supported exploration activities fueled by past experiences. However, these exploration capabilities did not mean that the team refused to adopt existing proven solutions. They did not have to reinvent the wheel and start from a blank page, when the organization already had a suitable solution available. This is what constitutes the contextual

[17] Castoriadis, Cornelius. *Devant la guerre*. Paris: Librairie générale française, 1983.

[18] Linhart, Robert, *L'Établi*, Paris, Minuit, 1978.

[19] Jullien, Bernard. "Relativiser le statut de la rupture dans la théorie évolutionniste. In *Approches évolutionnistes de la firme et de l'industrie*, Maurice Baslé, Robert Delorme, Jean-Louis Lemoigne, Bernard Paulré (coord.). Paris: L'Harmattan, 1999, 207–228.

ambidexterity[20] mentioned earlier: the designers' capability to choose the right rules to follow in each situation. In a project such as that of the Kwid, structural ambidexterity as sought by the CEO could turn out to produce very disappointing results if it did not, at the same time, leave enough freedom for teams to develop this contextual ambidexterity.

Conversely, the fact that the project was based on this intense use of Renault competencies meant that it was not easy for it to fulfill its ambition to become a permanent "Alliance project." The culture of challenging consensus, interdepartmental cooperation, and cost-saving cultivated by Renault for the Kwid design, the platform, and the powertrain had not been developed to the same degree by Nissan. Over time, Renault staff gradually took the lead on the platform and the powertrain, and developed their own vehicle. Nissan staff were left to focus on the upper body of the Redi-GO, for which they simply decided to reduce the number of added extras. To a large degree, this explains the difficulties that were experienced when the car had to emerge from a plant with Nissan standards, where it was new to have to manage initial production and final modifications simultaneously. Much more so than Renault, Nissan implemented the more mainstream strategy of splitting innovation into different parts without ever challenging its underlying tenets. While the process of inventing the development phase of innovation was a challenge even for Renault, it turned out to be extremely problematic for Nissan.

9.4 The Emergence of Peaceful Co-Existence Between the Entry Range and the Rest of the Segment

Since 2005, the title of Program Director at Renault has referred to the manager in charge of a group of products such as the Logan or Mégane ranges, along with all of their associated sister products. At this time the responsibility of project management was extended to both sales and marketing decisions as well as to all the products that were constructed on the same platform. The dominant development logic for a range is, to a certain degree, structural, as the milestones and the decision about when to freeze design are program specific and dependent on the level of experimentation present in the innovation process. The idea that the rules to be applied could differ slightly was not new to the Entry range, as other programs, such as the highly innovative program for electric vehicles, already partly deviated from those rules applied to the mainstream ranges.

[20] Gibson, Christina B., and Julian Birkinshaw. "The Antecedents, Consequences, and Mediating Role of Organizational Ambidexterity." *Academy of Management Journal* 47, no. 2 (2004): 209–226.

The "old" Entry range is to be revived in 2017 and 2018 and will subsequently be more clearly integrated into group policy. Speaking on this subject in the summer of 2016, Corporate Design Director Laurens van den Acker disclosed that, "For the next-generation Dacia, we will have to make the cars much more attractive and emotion laden, but we must do so without making them more expensive." At the same time, he stressed that, "Renault models will be increasingly distinct from Dacia's. They will have to be more 'Latin.' A third-generation Renault Logan will be identified with the other Renault cars of the range." Without completely returning to the fold, or becoming "Skoda-ized," as Gérard Detourbet put it, it was obvious that the program had become less exploratory and was more about continuing to tap into an opportunity that it had created for itself, and which no other competitor had succeeded in tackling so far. That meant systematically searching out carry-overs and carry-across to minimize cost impact at the same time as the cars were being made more attractive, as requested by Laurens van den Acker. This becomes primarily a question of exploiting the established design-to-cost rules rather than exploring radical new avenues, as might have initially been the case for the Logan. Conversely, during the same time period, other programs such as those involving segment D, which offer sophisticated driving-assistance systems to make the car truly autonomous, needed to become more exploratory.

The Kwid broke out of the box both geographically and organizationally by taking a shot at the emerging markets once again from the bottom up with a new platform and new motors and gear boxes. Yet, even when the CMF-A program was taking off in 2016, its Director was still not positioned in the same "organigramme" of those who managed the rest of the range. The Kwid now belonged to the Renault catalogue in a growing number of countries and was quickly requested by all the big regional bosses, including at the Renault Curitiba plant in Brazil. It was mainly Renault people who developed and built the platform, engines, and gearboxes, and they were more integrated into the company than were the members from the Logan team. As a result, while the Logan project was not of huge interest at the time of its development, as many were sure it would not work, the Kwid project was carefully monitored by all Renault functions, and each one contributed to its success.

There remains a persistent worry among Renault and Alliance executives that the company will become identified primarily with this type of product within the automotive industry. However, by 2016, it had become difficult to dispute the clear advantages of the approach and the competitive edge it gave the company. Similarly, the liberties that had been taken with the "Renault rules" to make the program succeed had also gained acceptance. They were considered justified by the need to explore new territories and manage procurement in emerging countries in which it was not realistic to demand that all decisions remain centralized at the head office. The altering of rules became acceptable even for products that would eventually be integrated into core programs of the company.

In addition, the environment in which the Kwid appeared was substantially different from that of the period in which the Logan was developed. During the phase in which the Logan and its associated brands were growing in importance, the European market had stalled, and Renault's range took a bigger beating than any other. Between 2007 and 2012, the Entry range went on to make up for the loss in volume and profits of the Clio, Mégane, and Laguna ranges combined. By contrast, the Kwid is not part of the European range, and it was launched just as the European market was recovering. With its new range spearheaded by the Captur, Renault outperformed the market and was a huge beneficiary of the recovery of the Italian and Spanish markets. In parallel, the negotiations for the company's entry into the Chinese market at the time did not feature the Entry range but focused instead on the category C and D SUVs, the Kadjar, and the Koleos. In France and in Europe, the new Espace and Talisman were seen as counteracting the impact of the Kwid and reassured those who feared that Renault was at risk of becoming perceived as a "low-cost" car manufacturer.

The Entry range and the sub-Entry range thus managed to position themselves in an increasingly central role within the company and its organizational processes. This became possible over time and in a gradual, rather than a fundamentally disruptive, manner. The process was nurtured by the long-standing development of the company and the accumulated resources that were already in place well before the launch of the Logan in the various functions of the company and built up over years in individual career paths. It emerged from the company's predisposition to ambidexterity that was present from top to bottom. It was thus possible, within a 15-year period, to roll out a profoundly original trickle-up approach in the automotive industry without simultaneously damaging the mainstream approach. Not only does this experience offer lessons in effective ambidexterity approaches, the process of gradual integration clearly highlights how strategies are most effective when they are both deliberate *and* emergent.[21,22] There is no doubt that the decision to develop the Kwid was the result of a correct analysis of the environment, which showed that there was a path to be followed by adapting the Entry range specifically for emerging markets. However, it was also the result of the steady maturing and expansion of the relevant skills and beliefs within the company, among people who were capable of paving the way. It almost goes without saying that this is what is at the heart of the success of the strategy that was adopted.

---

[21] Mintzberg, Henry, and James A. Waters. "Of Strategies, Deliberate and Emergent." *Strategic Management Journal* 6, no. 3 (1985): 257–272.

[22] Burgelman, Robert A., Webb McKinney, and Philip E. Meza. *Becoming Hewlett Packard: Why Strategic Leadership Matters*. New York: Oxford University Press, 2017.

Chapter Ten

The Global Innovation Playing Field

One of the key contemporary challenges for large groups is dealing with the playing field for innovation at a global level. Even if a large company is not necessarily in the best position to generate disruptive innovations, its competitive advantage should be, *a priori*, its ability to deploy innovation quickly on a worldwide scale. Consequently, the key capabilities of multinational groups include taking advantage of the specific features of geographical zones to target innovation strategies that would not work elsewhere, develop products adapted to these features, and quickly adapt these innovations to new contexts elsewhere.[1,2,3]

To put the new path of the global automotive industry initiated by the Kwid into context, it is useful to recall the challenges faced by internationalization in this sector and its recent acceleration for Renault. We then consider the relevance of the reverse-innovation concept in light of the fact that, while the Kwid has its

---

1 Bartlett, Christopher A., and Sumantra Ghoshal. "Managing Innovation in the Transnational Corporation." In *Managing the Global Firm,* Christopher A. Bartlett, Yves L. Doz, and Gunnar Hedlund (eds.). London: Routledge, 2011, pp. 215–255.

2 Prahalad, C. K, and Yves L. Doz. *The Multinational Mission: Balancing Local Demands and Global Vision.* New York: Free Press, 1999.

3 Ben Mahmoud-Jouini, Sihem, Florence Charue-Duboc, and Christophe Midler. *Management de l'innovation et Globalisation: Enjeux et pratiques contemporains.* Paris: Dunod, 2015.

roots in India, its design was based on mobilizing a global network of competencies. Finally, we look at the international deployment challenges faced by the Kwid and the CMF-A platform, which could complete the reversal of the innovation process.

10.1 Internationalization: The Hallmark of the Automotive Industry

Internationalization has certainly been a perennial feature of the automotive industry, especially during the various breakthroughs that marked its history.[4] When Henry Ford introduced mass production with the Model T, and with it the standardization of components and the assembly line, the innovation quickly took on an international dimension as Ford began assembling, and then producing, its cars on various continents[5]—including Europe, the birthplace of the automotive industry.[6] The change of industry leadership from Ford to General Motors, with what some would call "flexible production"[7] based on the multi-divisional organizational structure[8] of GM under the initiative of Alfred P. Sloan, also took that group into the international arena with the acquisitions of Opel (Germany), Vauxhall (England), and Holden (Australia).[9] The organizational innovation[10] variously known as "Toyotism," "Ohnism," "the Japanese model," and "lean production" in the 1980s–1990s was also confirmed by the process of establishing Japanese car manufacturers'

---

[4] Dassbach, Carl H. A. *Global Enterprises and the World Economy: Ford, General Motors, and IBM, the Emergence of the Transnational Enterprise.* New York and London: Garland, 1989.

[5] Wilkins, Mira, and Frank Ernest Hill. *American Business Abroad: Ford on Six Continents.* Detroit WI: Wayne State University Press, 1964.

[6] Bonin, Hubert, Yannick Lung, and Steven Tolliday. *Ford 1903–2003: The European History.* Paris: PLAGE, 2003.

[7] Hounshell, David A. *From the American System to Mass Production, 1800–1932: The Development of Manufacturing Technology in the United States.* Baltimore and London: The Johns Hopkins University Press, 1985.

[8] Chandler, Alfred Dupont. *Strategy and Structure: Chapters in the History of the Industrial Enterprise.* Cambridge, MIT Press, 1962.

[9] Bordenave, Gerard, and Yannick Lung. "Twin Strategies of Internationalization of the US Carmakers: GM and Ford." In *Globalization or Regionalization of the American and Asian Car Industry?* Michel Freyssenet, Koichi Shimizu, and Guiseppe Volpato (eds.). London: Palgrave, 2003, pp. 53–94.

[10] Boyer, Robert, and Michel Freyssenet. *The Productive Models: The Conditions of Profitability.* London: Palgrave, 2002.

plants overseas as part of the famous "transplants"[11] that followed a massive wave of exports.

This deployment corresponded to the start of the phase known as *globalization,* which was marked by an increased interdependence between industrialized economies and a relative convergence of markets, especially car markets. In response to the invasion of small, globally successful Japanese cars (such as the Honda Civic and Toyota Corolla), the Americans began their first attempts at designing global cars (GM Opel/Daewoo and Ford Escort), mainly for the North American and European markets. However, these attempts, which would be repeated, met with resistance from various locally developed engineering teams, which diverted the development paths (Ford Escort) or ended up creating products that would not travel (Ford Mondeo).

Today, we have entered a period of major restructuring in the global automotive industry. Alliances are made and unmade; mergers, takeovers, and demergers continue according to conditions, crises, and failures. The race for size has accelerated, as groups aim to take advantage of economies of scale and link the magic cost/quality dyad established by the Japanese. The minimum size of a global automotive group has grown from two million, to three million, to five million vehicles. To take into account the economies of scope resulting from the multiplication of models, strategic ambition has developed from the "world car" to the "world platform," which can enable the development of different vehicles on a shared technical underbody.[12] More recently, the car industry has moved toward more modular architecture, enabling greater flexibility while permitting savings by sharing sub-assemblies across several platforms and models adapted to different markets that retain their specific features, and convergence has remained relative.[13] The Kwid is part of this modular approach.

Initially focused on the industrialized world—the famous "Triad" of Kenichi Ohmae[14]—which continued to represent 84% of global vehicle production in 1990, the playing field of the global automotive industry has gradually shifted

[11] Boyer, Robert, Elise Charron, Ulrich Jergens, and Steven Tolliday, eds. *Between Imitation and Innovation: The Transfer and Hybridization of Productive Models in the International Automobile Industry.* Oxford: Oxford Univ. Press, 1998.

[12] Lung, Yannick, Jean-Jaques Chanaron, Takahiro Fujimoto, and Daniel Raff, eds. *Coping with Variety: Flexible Productive Systems for Product Variety in the Auto Industry.* Aldershot, Hampshire, England: Ashgate, 1999.

[13] Frigant, Vincent, and Bernard Jullien. "Comment la production modulaire transforme l'industrie automobile." *Revue d'économie industrielle,* no. 145 (2014): 11–44.

[14] Ohmae, Kenichi. *Triad Power. The Coming Shape of Global Competition.* New York: Free Press, 1985.

its focus to emerging markets.[15] From 2009, emerging countries accounted for more than half of the global vehicle production, and the trend is unlikely to reverse. Brazil, Russia, India, and China (BRICs) were the special targets, because their sales volumes generate economies of scale which are crucial in the car industry. The absence of such economies in car markets, which remained very volatile, probably accounted for the relative failure of the first world car designed for the emerging markets: Fiat's project 178, which engendered the Palio, the Siena, and their variants (including the Strada pickup truck).[16,17] Based on an idea originating from the corporate center (Italy) to use the Uno as a platform for global supply-chain management, the aim of the project was to exchange components and produce cars by connecting about 10 emerging countries (including all the BRICs). Even though these models met with some success, the project fell short of its initial ambition of an annual volume of one million vehicles. Paradoxically, this target was achieved by another car that was manufactured for the emerging markets and initially had no global ambition: the Logan, which produced a lineage with the Entry range.

10.2 The Entry Range: The Main Driver of Renault's Delayed Internationalization

Compared to this trend initiated by the Americans from the 1920s, and revived by the Japanese in the 1980s–1990s, European car manufacturers struggled to internationalize at the global level, in spite of local successes (Fiat in Brazil) as well as failures (Renault in the United States). It was not until the close of the 1990s that Renault took a decisive step in this direction with the Renault-Nissan Alliance and the formation of the Renault-Dacia-Samsung group after the buyout of Dacia in Romania and Samsung Motors in South Korea. The Alliance's takeover of Russian car manufacturer Avtovaz/Lada in 2012, and Mitsubishi in 2016 completed its global organization.

With five car models (Logan, Sandero, Duster, Lodgy, and Dokker)[18] now sold under the Dacia or Renault brands, depending on the countries, the

[15] Humphrey, John, Yveline Lecler, and Mario Sergio Salerno. *Global Strategies and Local Realities: The Auto Industry in Emerging Markets*. Houndmills, Basingstoke, Hampshire: Macmillan Press, 2000.

[16] Camuffo, Arnaldo, and Giuseppe Volpato. *Global Sourcing in the Automotive Supply Chain: The Case of Fiat Auto "Project 178" World Car*. International Motor Vehicle Programme, MIT Globalization Research, 2002, multigr.

[17] Dunford, Michael. "Globalization Failures in a Neo-Liberal World: The Case of FIAT Auto in the 1990s." *Geoforum* 40, no. 2 (2009): 145–157.

[18] From 2012, a new version of the Logan and Sandero was gradually introduced.

success of the Entry range[19] in emerging countries allowed Renault to become an intercontinental company: apart from Europe, Turkey, and South Korea, all the group's plants are predominantly devoted to manufacturing Entry-range cars (and sometimes other Renault models, of which a few hundred or thousand are made). Since 2016, Entry-range models are assembled in three South American countries: Brazil, Colombia, and Argentina. In Russia, apart from the Avtoframos plant in Moscow, an ambitious plan was made for the Togliatti site, where the Avtovaz plant produces derivatives of the Logan sold under the Renault and Lada brands. In Asia, the Logan continues to be produced, with some difficulty in Iran, while the Duster has allowed Renault to find its place in India. In Africa, three countries are involved: in 2008, Morocco witnessed the opening of the greenfield plant in Tangier, as well as the historical site of SOMACA in Casablanca,[20] Algeria and, at the other end of the continent, South Africa in a Nissan plant. Including the Pitesti complex in Romania that made the first Logans in 2004 brings the locations in which Entry-range cars are assembled to a total of 10 countries and 12 industrial sites. Practically nobody in the car industry can rival Renault in terms of deploying a range across four continents. With more than one million vehicles produced each year since 2013, Entry represented 45% of passenger car sales in the Renault group in 2015 (see Table 10.1); in the same year, the Duster, Sandero, and Logan all appeared in the top four models sold by the group globally, after the Clio IV.

This transformation of the French car maker into a global group also included design activities, which since 1998 have been consolidated at the Technocentre in Guyancourt—the largest hub of Renault activity,[21] with some 10,000 people. With the acquisition of Samsung, Renault gained access to high-level engineering resources capable of providing vehicles suitable for industrialized countries, gradually moving into making derivatives of Renault-badged models based on vehicles designed in South Korea, especially top-of-the-range cars such as the Talisman or Koleos. Moreover, Renault-Samsung completed the Technocentre team in Guyancourt by contributing to the Research, Advanced Studies, and Materials Engineering Department (DREAM).[22]

---

[19] Jullien, Bernard, Yannik Lung, and Christophe Midler. *The Logan Epic: New Trajectories for Innovation*. Paris: Dunod, 2013.

[20] Benabdejlil, Nadia, Yannick Lung, and Alain Piveteau. "L'émergence d'un pôle automobile à Tanger (Maroc)." *Cahiers du GREThA*, Univ. Bordeaux, no. 4 (2016). (To be published in *Critique économique*.)

[21] Bonnafous, Gilles. *The Renault Technocentre*. Paris: Hazan, 1998.

[22] Ben Mahmoud-Jouini, Sihem, Florence Charue-Duboc, and Christophe Midler. *Management de l'innovation et Globalisation: Enjeux et pratiques contemporains*. Paris: Dunod, 2015.

Table 10.1 Sales of the Renault-Dacia-Samsung Troup's Entry Range in 2015 (Passenger cars only)

Brands	Renault	Dacia	Total
Logan	222,974	101,256	324,230
Sandero	162,837	185,384	348,221
Duster	168,900	162,338	331,238
Lodgy	8,637	32,960	41,597
Dokker	18	29,530	29,548
Total Entry	563,366	511,468	1,074,834
Kwid	17,933		
Entry and Kwid	581,299	511,468	1,092,767
Group Total	1,822,965	511,510	2,414,503

(*Source:* Atlas Renault 2015.)

The case of Dacia was more complex. When Renault took over the Romanian car maker in 1999, any engineering resources were disconnected from modern competencies and technologies. Their gradual upgrading and the development of competencies in Romania led to the creation of Renault Technology Romania (RTR), which, despite failing to achieve autonomy, plays an increasingly predominant role in the development activities of the Entry range.

Finally, the fourth American Technocentre is in Brazil. It started its activity on the basis of the traditional "tropicalization" approach, associated with the strategy sometimes termed "glocalization"—that is, adapting vehicles designed at the corporate center to local conditions, especially to meet customer expectations (the clearest example being the flexfuel engine, which is specific to Brazil). Again, the development of competencies would result in Brazil's taking the lead on derivative versions of the Entry range: the design of the Sandero Stepway, a sporty version of the Sandero (R.S. 2.0), and the Duster pickup (Oroch). Although it does not have Technocentre status, despite proven competencies, India became the fifth global design center of the group with the *ex nihilo* development of the Kwid by the 2ASDU team.

10.3 The Kwid: A Case of Reverse Innovation?

It has already been stressed several times that the Kwid design was a major breakthrough in the organization of car design, for Renault as much as for the global automotive industry. Frugal/fractal engineering methods made it possible to reduce the production costs of a modern car by half, when compared to the reference model of the Entry (which had itself reduced the costs by half). Therefore, it involved an innovation that targeted the bottom of the range (the base of the pyramid, as defined by Prahalad[23]), approaching disruptive innovation as defined by Christensen[24]—except that the bottom of the range was attacked by an *insider,* not a new entrant, and the Kwid is located farther down the pyramid than the Logan.

The Kwid was the first Renault car that had not been designed at the corporate center, but at a local project war room in a foreign country. There is a clear reversal of dynamics with regard to the dominant strategy described by "product cycle" theory,[25,26] which assumes that innovative products are designed in the key countries (the United States, for Vernon) before being spread to other industrialized economies (Europe), and finally arriving in developing countries at the end of the cycle. For the automotive industry, the production transfers of the Volkswagen Beetle to Mexico and the Citroën 2CV to Portugal at these cars' end of life—not forgetting the Renault 12 in Romania, which provided the Dacia models for several decades—illustrated this pattern perfectly. For southern countries, the obsolete models were retained by ultimately phasing out written-off production tools from central regions.

In the first phase of the recent globalization, the emerging countries became testing grounds for innovative products designed in the corporate center. Although designed centrally, products were industrialized on the outskirts to take advantage of the lower wages or to experiment with new forms of organization by benefiting from the flexibility of these spaces in case of failure. The production of the Audi TT in Hungary, or the Volkswagen New Beetle in Mexico, to supply the markets in industrialized countries illustrated this approach.[27] We might also cite the first attempts at modular production

[23] Prahalad, C. K. *The Fortune at the Bottom of the Pyramid: Eradicating Poverty Through Profits.* Upper Saddle River, NJ: Wharton School Publishing, 2005.

[24] Christensen, Clayton M., Michael E. Raynor, and Rory McDonald. "What Is Disruptive Innovation?" *Harvard Business Review* 93, no. 12 (2015): 44–53.

[25] Vernon, Raymond. "International Investment and International Trade in the Product Cycle." *The Quarterly Journal of Economics* 80, no. 2 (1966): 190–207.

[26] Lung, Yannick. "Repenser les trajectoires de la géographie de l'innovation." In *Les trajectoires de l'innovation.* C. Bouneau and Y. Lung (dir). Bruxelles, Belgique: Peter Lang, 2014, pp. 201–223.

[27] Layan, Jean-Bernard. "L'innovation péricentrale dans l'industrie automobile: une gestion territoriale du risque de résistance au changement." *Flux,* no. 63–64 (2006): 42–53.

in Brazil, which is another form of innovation by the automotive industry in emerging countries.[28]

The originality of the Kwid with regard to the car industry lay in the design of the vehicle made in India with a global ambition. The car industry did not, of course, initiate what the literature calls "reverse innovation." The concept was introduced in a 2009 article in the *Harvard Business Review*,[29] which described the design of new medical devices in China and India by General Electric for these markets first, achieving significant cost reductions through frugal engineering methods, but finding sales in the United States since then.

One of the creators of this concept, Vijay Govindarajan,[30,31] highlighted the characteristics of reverse innovation. It did not target the top of the income pyramid, corresponding to expensive products (such as upmarket cars), at which innovations would be introduced before trickling down to lower segments. This is the traditional pattern, but now the social background is the middle classes in emerging countries: lower-income households who nevertheless have some spending power. So the reverse innovation is designed on site, combining engineering and marketing, to dive into the expectations of the local market as well as into the local supplier network. Finally, it preserves the main features expected by customers while excluding those marginal features that are hardly used but increase the price.

Numerous other examples of this reversal of the usual deployment paths[32] have been documented (such as mobile-phone payments in Africa), and multiple academic contributions have sought to examine the concept of reverse innovation in greater detail, aiming to increase its precision beyond the intuitive resonance that probably accounted for its initial success.

---

[28] Lung, Yannick, Mario S. Salerno, Mauro Zilbovicius, Ana Valéria Carneiro Dias. "Flexibility Through Modularity: Experimentations of Fractal Production in Europe and Brazil." In *Coping with Variety: Flexible Productive Systems for Product Variety in the Auto Industry,* Yannick Lung, Jean-Jaques Chanaron, Takahiro Fujimoto, and Daniel Raff (eds.). Aldershot, Hampshire, England: Ashgate, 1999, pp. 224–257.

[29] Immelt, Jeffrey R., Vijay Govindarajan, and Chris Trimble. "How GE Is Disrupting Itself." *Harvard Business Review* 87, no. 10 (2009): 56–65.

[30] Govindarajan, Vijay. "Conversations: Reverse Innovation: An Interview with Vijay Govindarajan." *Research-Technology Management* 55, no. 6 (2012).

[31] Govindarajan, Vijay, and Chris Trimble. *Reverse Innovation: Create Far from Home, Win Everywhere*. Boston, MA: Harvard Business Review Press, 2012.

[32] Based on the analysis of innovation deployment processes in multinational groups (refer to S. Ben Mahmoud-Jouini, et al. *Management de l'innovation et Globalisation: Enjeux et pratiques contemporains, op. cit.,* especially case studies, pp. 31–53 for Orange or pp. 56–88 for Air Liquide).

Some have tried to articulate it with innovation terms such as "frugal" (*Jugaad*), "disruptive," "open," etc. This is especially so in the case of Zeschky and his co-authors,[33] who differentiate the various types of innovation in emerging countries. One type includes innovations focused on costs, which do not modify product features but instead endeavor to drastically reduce the production costs in relation to competitors (such as the batteries designed by the Chinese manufacturer BYD for its electric car). Then there are "good-enough" innovations, which adapt (or, in fact, reduce) features to the local expectations of emerging markets while cutting costs. Finally, there are the frugal innovations, based on an original architecture that offers new features at a lower cost. The Kwid could be classified in any of these three categories: it is a car like others (case 1); but features that are not essential on the Indian market, such as ABS, ESP, etc., are stripped away (case 2); and it is also a new architecture for the Alliance, the Common Module Family, designed according to frugal engineering methods (case 3). For these authors, the concept of reverse innovation is cross-functional, since it describes the capacity of these innovations (whether type 1, 2, or 3) to be distributed on a global scale—especially to rich countries, if the hypothesis of Govindarajan and Ramamurti is maintained.[34] Regarding the Kwid, this raises the question of whether it is possible for India to be a "lead market" in this segment of the car market, since the special features of the Indian market weighed so heavily on the design of the vehicle and hence, *in fine,* the platform and modules (Cf. Chapter 8).

Refining the reverse-innovation concept,[35] others have sought to examine it in detail by breaking down innovation into various phases and specifying the place, nature, extent, and forms of reverse-innovation integration, along with the relationships connecting these different dimensions.[36] In this context, von Zedtwitz et al.[37] proposed specifying the stage of the process at which innovation is reversed by differentiating the phases when ideas are generated, products developed, and products adopted. They characterized the extent of

---

[33] Zeschky, Marco B., S. Stephan Winterhalter, and Oliver Gassmann. "From Cost to Frugal and Reverse Innovation: Mapping the Field and Implications for Global Competitiveness." *Research-Technology Management* 57, no. 4 (2014): 1–8.

[34] Govindarajan, Vijay, and Ravi Ramamurti. "Reverse Innovation, Emerging Markets, and Global Strategy." *Global Strategy Journal* 1, no. 3–4 (2011): 191–205.

[35] Radojévic, Nebojša. "Reverse Innovation Reconceptualised: Much Geo-Economic Ado about Primary Market Shift." *Management International* 19, no. 4 (2015): 708–782.

[36] Hussler, Caroline, and Thierry Burger-Helmchen. "Inversée vous avez dit inversée? Une typologie stratégique de l'innovation inversée." *Revue française de gestion* 42, no. 255 (2016): 105–119.

[37] Zedtwitz, Max Von, Simone Corsi, Peder Veng Søberg, and Romeo Frega. "A Typology of Reverse Innovation." *Journal of Product Innovation Management* 32, no. 1 (2014): 12–28.

reversal based on the number of stages in the innovation process that are reversed. Low-reverse innovations are those in which only one stage witnesses a reversal of its geography, while high-reverse innovations are those in which the phases of idea generation, product development, and product adoption lead to geographical "comings and goings" around the world. As analyzed by Laperche and Lefebvre,[38] the development of the Logan fell outside the scope of reverse innovation—unless the low-end version was taken into account—because the Logan had only relocated its sales targets and manufacturing. It kept the design very close to the corporate version and favored the carry-over. Other lineage products (the Sandero and, especially, the Duster) had granted the Brazilian and Romanian design offices more power and created more radical conditions in the reverse-innovation approach. The Kwid was born within the framework of the 2ASDU, an Alliance structure that managed a project war room in India that was responsible for developing a platform, engine, gearbox, and two new products. There had been a clear transition from weak reverse innovation to strong.

On the issue of the source of value, the local engineering teams' ability to propose an alternative to the ideas of corporate engineering apparently indicated a fuzzier distinction between the Logan and Kwid, for two reasons. First, the Entry program initiated by the Logan had already started, from Paris, the main part of the breakthrough process with the standard strategy and had made an alternative proposal around the concept of a "bare minimum." The Kwid radicalized and relocalized this approach in India, expanding it from the issue of costs to encompass the issue of *performance*—that is, the functional characteristics of the vehicle and, hence, the requirements allocated to the components and the interactions that would develop between them.

The issue of the style was initiated with a "green" sketch from the Mumbai studio. "We have played our part by challenging the paths developed at the Technocentre, using our knowledge of the Indian automotive market," recalls Jean-Philippe Salar, manager of the Mumbai studio at the time. "To stress the importance of an attractive style as our primary USP over others was really a certainty for us, and this was shared by those who knew India." In the first part, we noted that resident engineering teams' control over which key sales arguments (USPs) were selected, as well as how they were developed under cost constraints, had greatly increased from the Logan to the Kwid. At the same

---

[38] Laperche, Blandine, and Gilliane Lefebvre. "The Globalization of Research & Development in Industrial Corporations: Toward 'Reverse Innovation'? The Cases of General Electric and Renault." *Journal of Innovation Economics & Management* 2, no. 10 (2012): 53–79.

time, it allowed the radicalization of the "design-to-cost" principle as well as much greater freedom in defining the bare minimum.

From the Logan to the Kwid, there was a kind of learning involved with regard to reverse innovation. It was not—or at least, not only—a matter of the degree of decentralization and/or the weight given to local engineering teams in emerging countries. Based on what the literature on ambidexterity consistently revealed, this issue was coupled with the problem of integrating central and local competencies and structures: the Logan, and also the Kwid, still leant heavily on Corporate central resources. The project team was supervised, and the lessons learned at the head office provided the competencies that made intrusive management and fractal innovation possible.[39] Moreover, when standards were called into question, what was most important was the fact that they were present, and, when reassessment led to redefinition, this redefinition was submitted to corporate once more. This was the condition on which innovation risks were managed, and on which the innovation designed in India was able to travel to other destinations and—possibly—developed countries. To a certain extent, reverse innovation was based as much on the ability of project structures and resident engineering teams to absorb the corporate competencies and redirect them to "their projects" as on Corporate's ability to transform what was created by the authors of reverse innovations into alternative methods or rules (see Chapter 8).

10.4 The Role of Proximity

Unlike the modular architecture, the bricks of which could be independently designed and made using the set of interfaces, and in spite of the automotive industry's efforts, a car remains a fully integrated architectural product, insofar as the overall consistency of the project should take priority over the development of each of its components.[40] Numerous interactions were required almost daily between the various functions to align the project with a product. This mobilized the various means of communication to share the tacit knowledge of technical experts.

Thus, as noted, for more than two decades the car industry adopted the principle of colocating the teams in a project war room in which various functions

---

[39] Jullien, Bernard, Yannick Lung, and Christophe Midler. "De Logan à Kwid. Ambidextrie, innovation inversée et fractale, design-to-cost: les recettes de la stratégie Entry de Renault." *Cahiers du GREThA,* Univ. Bordeaux no. 19 (2016).

[40] Fujimoto, Takahiro, and Youngwon Park. "Complexity and Control: Comparative Study of Automobiles and Electronic Products." *MMRC Discussion Paper Series,* no. 352 (2011).

could meet continually.[41] The 2ASDU project war room was a place in which senior managers (vehicle project manager, vehicle architect, and powertrain project manager, as well as the engine and transmission managers, customer requirement managers, sales and marketing managers, purchasing managers, industrial styling managers,[42] etc.) and their direct associates at the same level gathered together. As the managers were experienced expatriates, the Indian engineers and technicians mostly had to fully design a brand-new vehicle for the first time (see Chapter 8). This colocation facilitated the knowledge transfer.

However, this organized proximity[43,44] was more than merely physical. The initial intention was to have a mixed Renault-Nissan team, but the team that ultimately took responsibility for the design of the vehicle and platform gradually came to be composed mainly of Renault expatriates. The Nissan expatriates concentrated on developing the upper body of the I2, leaving the Renault expatriates to deal with the basic development of the vehicle. Simply having the same workplace is not enough to build a team: thought processes, standards, and values also have to be shared. Yet Renault has a shared professional culture that has often been improved by prior experiences on the Entry range, while Nissan has its own culture. Therefore, a kind of division of labor had been established, which the modular strategy at the source of the CMF-A favored. This proximity also had to be developed between Renault and Nissan, including through technical mediation artifacts. This sometimes had to rest on some "bricolage": in spite of more than 10 years of the Alliance's existence and the significant effort to achieve convergence in terms of production systems, the two groups' information systems remained separate and incompatible. This led to the creation of a large Excel table to share components between I2 and XBA.

The team members in the Indian project war room were completely engrossed in the project, and this was certainly an important factor in the success of the design. Sharing the conditions of use of the vehicle on the Indian roads and cities on a daily basis, watching the movement of cars that corporate considered "impossible to sell" but which were among the most popular models on the Indian market—such direct observations would lead to a change in the

[41] Midler, Christophe. *L'auto qui n'existait pas. Management des projets et transformation de l'entreprise*. InterEditions, Paris, 1993. Nouvelle édition Dunod, Paris, 2012.

[42] In fact, the Design Department is in another RNTBCI building, but still on the same site.

[43] Dalmasso, Cedric, and Rémi Maniak. "La genèse d'un centre de R&D à l'international. Le cas de l'industrie automobile." *Management International* 19, no. 4 (2015): 83–94.

[44] Carrincazeaux, Christophe, and Yannick Lung. "La proximité dans l'organisation de la conception des produits automobiles." In *Proximité(s): approche interdisciplinaire,* M. Bellet, T. Kirat and C. Largeron (eds.). Paris: Hermès, 1998, 241–265.

perception of the market and customer expectations with regard to a development that would be carried out at the Technocentre in the Paris suburbs. This facilitated exchanges with the marketing department and distribution networks, which are located in New Delhi.

Moreover, the short distance (half an hour by car) to the assembly plant promoted exchanges and interactions between engineering and manufacturing. This was also the site at which the new 2ASDU powertrain plant was built to produce the engines and transmissions that would equip the Kwid and Redi-GO. The exchanges and interactions were especially intense during the frugal-engineering and product-development phases.

Finally, this local organization was concerned with more than internal issues, because it also involved suppliers who produced more than 80% of the vehicle's value. As was stressed in Part I, this meant achieving targets in terms of production cost and quality through regular, sometimes daily, interactions to work on the design, process, quality, etc. Although the Alliance plant did not have a suppliers' park in the immediate vicinity, as was widespread in the car industry,[45] the location of close to two-thirds of the Kwid suppliers in the Chennai area greatly facilitated these exchanges and contributed to success.

Even so, the new 2ASDU project war room did not have all the resources and competencies required to handle all the design stages of a new platform, two vehicles, and a powertrain. As noted in the first part, there were no heavy technical facilities to carry out tests in India, nor were there very specialized technical competencies to solve the problems encountered during the product development phase (such as noise). Therefore, the office regularly requested the central facilities of the Renault or Nissan group, whether in France, South Korea, or Japan. The exchanges were facilitated by transient proximity (mobility of the engineers), and maintaining relations was easier remotely within the framework of codified and standardized procedures, involving less of the tacit knowledge that is, on the other hand, essential for producing new knowledge. Again, it was noted that knowledge of interpersonal skills (social proximity) often promoted the remote mobilization of central resources on the project (see Part I).

If we add the role of the Cooperative Laboratory for Innovation (LCI) at the start of the project, it would seem that the design stage had been organized through a network focused on India, where the core of the development was carried out in the 2ASDU project war room, mobilizing design or central-engineering competencies in industrialized countries and organizing/improving

---

[45] Frigant, Vincent, and Yannick Lung. "Geographical Proximity and Supplying Relationships in Modular Production." *International Journal of Urban and Regional Research* 26, no. 4 (2002): 742–755.

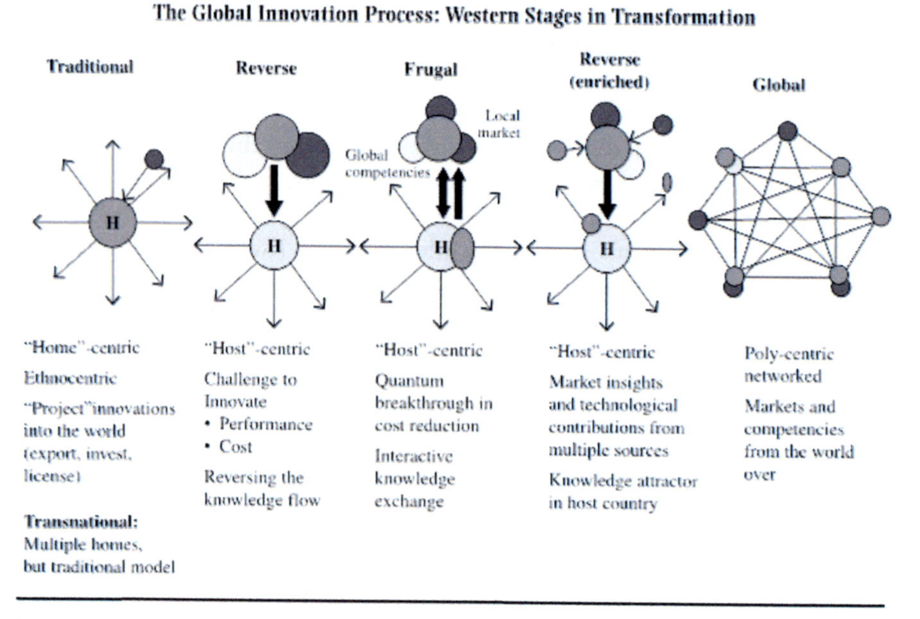

Figure 10.1 Various configurations of the overall organization of innovation.

the local competencies of suppliers. In relation to the analysis of the transformation of global innovation management patterns proposed by Yves Doz[46] (see Figure 10.1), we are approaching the *enriched reverse innovation* model, but with two slight differences: the multipolarity of the "corporate" for the Alliance, and questions regarding the feedback of this innovation, which is the Kwid, in the central markets.

10.5 Will India Be a Lead Market for the Kwid?

Of course, the Kwid was not the first car to be designed in India. Local manufacturers such as Maruti Suzuki[47] and Tata have competencies in this field. These Indian cars, such as Tata's Indica, enjoyed significant local success in both India and in neighboring markets, but they were too specific to be sold in other parts of the world. Some analyzed the Nano, with its €2,000 price tag—a

[46] Doz, Yves L., in Ben Mahmoud-Jouini, Sihem, Thierry Burger-Helmchen, Florence Charue-Duboc, and Yves L. Doz. "Global Organization of Innovation Process." *Management International* 19, no. 4 (2015): 113.

[47] Bhargava, Ravinda Chandra. *The Maruti Story: How a Public Sector Company Put India on Wheels*. Noida, India: Collins Business, 2010.

model of frugal engineering—as a disruptive innovation for the global automotive industry.[48,49] But it did not have the desired international distribution, and even in India its sales are now under 1,000 units per month. Its technical problems and its weak performance compared to competitors' models defeated its ambition.

From then on, the incontestable success of the Kwid was followed by that of the Redi-GO, as the first weeks of the Datsun car's sales suggested. Could this success be reproduced in other parts of the world for a car that had been broadly conceived for the Indian market? In other words, could India be a market in which the innovation appeared before being diffused—a "lead market"[50] for small cars, as theorized by Marian Beise?[51,52]

Romania, in fact, had been such a lead market for the Entry range because, even if the Logan had been conceived at Corporate (in France), the starting point of its design had been the characteristics expected by the Romanian market. If the Logan found real success in Romania—albeit less in the Central and Western European countries that had also been targets—the breakthrough came from its reception in the main European car markets (especially France and Germany) and the emerging countries. Such a reception had not been anticipated at the outset.

Given the volumes involved in justifying the investment, the Alliance's real challenge lay in India's becoming a lead market for the emerging countries. The car's reception in Brazil would be the acid test. Uncertainty persisted over whether a complete reversal would take place—that is, whether, like the Logan, the Kwid would be sold in industrialized countries, starting with Europe. Since 2015, the European automotive press (especially in France) had repeatedly announced its possible marketing from the Tangier site, where exploratory studies planned its allocation. Automotive journalists who tried out the car became champions of its going on sale in Europe. However, it was unlikely to happen for a while, because there were still steps to be taken in terms of production costs

---

[48] Freiberg, Kevin, Jackie Freiberg, and Dain Dunston. *Nanovation: How a Little Car Can Teach the World to Think Big and Act Bold.* Nashville, TN: Thomas Nelson, 2011.

[49] Chacko, Philip, Christabelle Noronha, and Sujata Agrawal. *Small Wonder: The Making of the Nano.* Chennai, India: Westland, 2010.

[50] Tiwari, Rajnish, and Cornelius Herstatt. *Aiming Big with Small Cars: Emergence of a Lead Market in India.* India Studies in Business and Economics: Springer International Publishing, 2014.

[51] Beise, Marian. *Lead Markets: Country-Specific Success Factors of the Global Diffusion of Innovations.* Heidelberg: Springer, 2001.

[52] Beise, Marian. "Lead Markets: Country-Specific Success Factors of the Global Diffusion of Innovations." *Research Policy* 33, no. 6–7 (2004): 997–1018.

Table 10.2 Change in the Average Characteristics of Cars Sold in Europe

Characteristics	2001	2014
Price (€)	20,060	26,435
Power (kw)	74	90
Weight (kg)	1,696	1,859
Length (m)	4.15	4.28
Width (m)	1.71	1.78

(*Source:* International Council on Clean Transportation, Pocket Book, 2015.)

so that the Kwid would not eat into the sales of the other Dacia or Renault models, and there were probably vehicle quality issues to be dealt with as well. The Kwid was particularly frugal to meet the minimalist expectations of the Indian market, especially in terms of safety. That contrasted with recent developments in the European market where, owing to safety and environmental regulations, the trend was to make the vehicle heavier and increase its price (see Table 10.2).

Therefore, the Kwid needed to be enriched and "taken up a notch"—that is, made heavy enough to travel. This was noted earlier with regard to Brazil, which was the first step. Note that the price could not be the €3,500 proposed for the basic model in India, taking into account the regulations, especially in terms of safety and environment. The basic expectations for cars were also different: power windows, ABS, ESP, airbags, the strengthening of the architecture, and other modifications to increase the weight and hence the price by reducing energy performance. Overall, it seemed unlikely that the price would come in under €6-7,000. But even that would be nearly 25% lower than the basic price of the Entry. This would increase access to the new car market with a new customer base.

Perhaps the Kwid would not end up in the markets of the industrialized countries. But even if it did not, other CMF-A derivative models that took their first steps in South India, in the 2ASDU project war room, probably will. The modular design of the Kwid means that it should provide a basis for a range of new products in the years to come.

Conclusion

For several years, many writers have emphasized the advantage of company strategies focused on developing products adapted to the most financially constrained customers. Christensen's trend analysis of disruptions[1] demonstrates the risks involved in pursuing *sustaining* strategies for developing increasingly sophisticated and costly products. The works of Vijay Govindarajan and Chris Temple on reverse innovation,[2] and, previously, *The Fortune at the Bottom of the Pyramid* by Christopher K. Prahalad[3] highlight the economic importance of potential markets in high-growth emerging countries—so much so that even today's large globalized companies from industrialized countries could try to invest in these markets by developing innovations adapted to them.

This is also a call for the frugal innovation approach or *jugaad*—a concept articulated by Navi Radjou, Jaideep Prabhu, and Simone Ahuja in their book *Jugaad Innovation*[4]—which advocates the use of ingenious improvised solutions like those we often see in developing countries. Today, this management trend is rich and diverse. However, if we were to sum it up in one sentence, we could

1 Christensen, Clayton M. *The Innovator's Dilemma: When New Technologies Cause Great Firms to Fail.* Cambridge, MA: Harvard Business School Press, 1997.

2 Govindarajan, Vijay, and Chris Trimble. *Reverse Innovation: Create Far from Home, Win Everywhere.* Boston, MA: Harvard Business Review Press, 2012.

3 Prahalad, C. K. *The Fortune at the Bottom of the Pyramid: Eradicating Poverty Through Profits.* Upper Saddle River, NJ: Wharton School Publishing, 2005.

4 Radjou, Navi, Jaideep C. Prabhu, and Simone Ahuja. *Jugaad Innovation: Think Frugal, Be Flexible, Generate Breakthrough Growth.* San Francisco: Jossey-Bass, 2012.

say that its central idea is that it might be better to meet the needs of the massive population of existing "non-customers" who cannot afford increasingly sophisticated and costly products, rather than perpetually competing in demanding markets that are saturated with the wealthiest customers. The Jugaad unifying mission, then, would be to "find ways to convert non-customers into customers."

Nevertheless, the number of large companies that have actually ventured into such strategies is negligible. Mostly, their initiatives involve a handful of isolated operations that are often blown out of all proportion in corporate social responsibility reports. To move beyond the exhortations of academics or consultants and find the real truth behind the inspirational words, we must hammer out the uncertainties of *how* such strategies can be implemented in an economically sustainable way. The story we have told in this book is an iconic example that provides a wealth of information on this uncertainty of "how," across five different levels.

First, it illustrates the realities of international markets at the lower end of the distribution: in this case, Indian vehicles under four meters in length and less than 3,5 lakhs in cost (€4,500), which represent 70% of vehicles sold in India. It also shows repeated failures to penetrate the Indian market with products derived from those designed for mature markets, stripped down to give them a chance of meeting price demands. However, these failures did not lead to a dead end, since the Kwid arrived a few months later, challenging the segment leader, Maruti's Alto. This affirms that strategies that neglect the need to design products around the specific expectations of customers are bound to fail.

Moreover, our case illustrates the sheer magnitude of the collective engineering skills needed to design complex mass-produced products when frugality is paramount. As Navi Radjou notes, it is not easy to be ingenious when the product to be designed contains over 1,000 components that must all meet stringent quality and safety standards; when it will be built with processes capable of producing a car a minute; when design is a collective effort of several hundred people, mobilizing engineering centers around the world; and when several hundred partner suppliers are involved (over and above the companies of the Alliance).

We have seen that the spectacular success of this "design-to-cost" approach (bearing in mind that the Kwid's total cost is around half that of the Sandero, which was itself unrivalled in Renault's range) resulted from a unique approach that we call *fractal innovation*. Our story contains no spectacular breakthroughs, like Dyson's bagless vacuum cleaner. Nor is it a tale of Japanese *kaizen*, with continuous and incremental developments improving the existing situation step by step. Instead, our case is all about analytical capability and systematic questioning of the available solutions for *all* action variables at *all* levels. These action variables included technical solutions for product and process, procurement, distribution, and sales and marketing. The levels involved ranged from the overall style of the product and unique industrial alternatives such as open plants, right down to details such as the spacing of cabling fastenings, the number of wheel nuts, or selecting screwdrivers for the assembly line. All these questions were targeted obsessively, driven by the purpose of the project—which was partly frugality, of course, but also took account of local factors such as India's unique driving conditions.

We have seen how this fractal innovation flowed not only from the skills of the engineers who designed the platform, but also from the organization's structure and the general management style of the program we called *intrusive management*: because the project was so unique, managers did not shy away from challenging the original grounds for corporate design standards at the micro level. This gave the project an autonomy without which the countless transgressions we observed would not have been chased down.

Our story represents an important development in the dynamics of streamlining design in large automotive groups that have prevailed since the mid-1990s. These dynamics were aimed at accelerating and multiplying developments in new products, motivated (quite rightly) by the need to eliminate quality risks and reduce development cost constraints. Gradually, innovation deserted the field of development engineering and was transferred upstream, winding up in advanced engineering and exploration cells.[5] The concept that led to the Kwid also originated in such a cell, the Cooperative Innovation Laboratory (LCI).

Today, development tends to be reduced to "cut-and-paste" assembly of standard solutions that have been proven in previous vehicles or external studies (although it had always been this way to some extent, purely due to the myriad constraints and inherent complexity of the product). The project led by 2ASDU reinvented engineering as a creative means to transcend this industrialization of car development. This is a key issue in an increasingly unstable and uncertain competitive world, giving the project an edge in terms of the capability to respond and adapt.

The third lesson is the global stage on which the Kwid story unfolded. The previous point could suggest a return to the past, to the revival of quintessential car projects from the 1960s to the 1980s, from the DS to the Twingo. Nothing could be further from the truth. These projects were a part of a corporate culture deeply rooted in the philosophy of the company's headquarters; the processes for defining new products were deployed very close to the strategic, technologic, and commercial core of the company. At that time, everyone from the Chairman to the engineer knew they were making a car for the married couple, the family, the folks next door. This was no longer the case with the Logan and the Kwid.

5 Midler, Christophe, Rémi Maniak, and Romain Beaume. *Réenchanter l'industrie par l'innovation*. Dunod, Paris, 2012.

Today, we are witnessing two kinds of internationalization. On the one hand, development is trying to understand and adapt itself to the mobility needs of customers who have nothing in common with the locals of (for Renault) Boulogne-Billancourt or Guyancourt. At the same time, development processes are being deployed on a global scale. Clearly, the two movements are not independent. It was no surprise that the Kwid's style was popular in India, given that its initial sketch was proposed by a designer from Mumbai. We saw how far the project pushed the analysis of competitor vehicles in the Indian market; how it was inspired by industrial solutions from local suppliers. We saw how the base in Chennai brought together Indian engineers from the local subsidiary of the Alliance.

However, this "Indianization" was not brought about single-handedly by the project war room in Chennai; it was only possible because of the core members of the team, comprising expatriates who brought skills acquired specifically from the Entry program (the Logan and its derivatives), and collaboration between the Technocentre and Romania. When necessary, which was often, the project could also turn to the "back offices" of technical expertise of the parent companies in France, Japan, and sometimes South Korea. The internationalization processes of the companies were merged in the 1980s and 1990s in terms of production and sales. Today, internationalization is centered on the field of design, with new challenges in managing these multipolar design "footprints": What are the specific roles of the new offshore design units? Who coordinates them, and how? These questions had been asked for the Entry and are again being asked for the new platform, with the Kwid being developed in Brazil.

The fourth lesson focuses on project management in an alliance. Alliances between groups have been one of the key themes in globalized industries for more than 20 years. The pursuit of scale effect and the magic word "synergy" inspired these mergers, which *a posteriori* turned out to be less profitable than on paper. The Renault-Nissan Alliance, formed in 1999, is a rare example of a long-term success. Realizing the much-vaunted synergies depends on two very distinct principles: complementarity and integration.

Complementarity is the allocation of areas of operation between the partners such that everything is encapsulated in an optimal way, and redundancy is eliminated. Founded on the principle of division of labor, this type of cooperation infringes least on the identity of the companies, although it may create limitations and frustrations. Thus, based on their initial sales and marketing situation, Renault and Nissan quickly divided up their international operations: for example, Nissan was given leadership in the US and China; Renault in Europe, South America, and Africa. For engine engineering, Nissan was given petrol engines; Renault diesel engines.

The other cooperative strategy is the integration of activities. Within the Renault Nissan Alliance, this took place very quickly in supplying, delivering scale effects, and boosting the commercial advantage over suppliers to obtain an obvious and rapid competitive edge for the Alliance. With regard to vehicle engineering, responses were more ambiguous and complex to implement. While the principle of a common platform looked good in theory, its application was difficult. The vehicle launch cycles of Renault's and Nissan's product plans did not generally coincide, so the rule adopted was to delegate the development of the platform to the company that would be the first to launch the product, before the other took it over to develop its own product on it. In practice, the rule was challenging to apply, due to the profound contrasts between the professional engineering cultures, in both product and techniques. To "take over" another manufacturer's technical solution meant understanding its internal philosophy and being able to evaluate it in terms of one's own technical criteria. In practical terms, that came down to repeating the work of the first designer—and, in general, significantly altering the result.

From this point of view, the 2ASDU story was also revolutionary: for the first time, a "single Alliance platform, two simultaneous products" principle had been essayed. This brought to light not only the difficulties of the exercise, but also the advantage of mediating the two companies by driving the projects jointly.

With regard to the difficulties, we have seen how the asymmetry between Renault's and Nissan's involvement weighed on the project. This asymmetry could be easily explained by lessons Renault had already learnt from the Entry, whereas Nissan's identity did not affect the company's decision to invest itself heavily in a low-end strategy. Besides the differences in strategic focus, we have also seen how the practical implementation of the "one Alliance platform, two simultaneous products" principle, particularly by Nissan engineering, required the intervention of the Program Director to prevent the formula remaining a hypothesis in a PowerPoint presentation. Splitting activities was one thing; actually integrating design activities was quite another.

Even though the process was difficult, we must also remember that its success validated a unique strategy of integration, based on the principle of a unique mediation between the two companies. One example was the decision-making autonomy of the program management, under the final authority of the CEO Carlos Ghosn, so that decisions could be adapted to a context specific to the project. In contrast to a functional department integration process, in which the rules of the two companies have to be compared and reconciled, here the project had the responsibility and scope to decide what was appropriate to its context. This integration strategy may be modest in ambition, but it is certainly more effective than the delegating/reworking strategy we have observed for other vehicles. Above all, it ensures that decisions are based not on the political balance of power within the two companies, but on the relevance of these decisions to the project's specific context—and, most crucially, the expectations of target customers.

The last learning was at a strategic level. It proved how an improbable idea could be developed and confirmed in a large global group, while simultaneously cultivating the factors that made it feasible. When Louis Schweitzer floated his idea of a $5,000 car to Renault in 1997, he was surrounded by skeptics on all sides. After three years of tentative exploration, the initial act of this emerging strategy was modest and cautious. The Logan project went ahead with minimal investment and set itself an initial sales target that was ludicrously low for the car industry (60,000 vehicles), guaranteeing meager profits. Thirteen years later, Carlos Ghosn's decision to support products like Renault's Kwid and Datsun's Redi-GO was a key indicator of the maturity of such a strategy: significant investments in a new powertrain, high-profile launches of two distinct products, ambitious volumes. It was a strategy that took a long time to emerge and that long retained an air of minimalism, of being the exception that proved the rule—in spite of its obvious financial success and significant sales (Entry represented 45% of Renault group's volumes in 2015). Between the two, a collective capability of frugal engineering, global entrepreneurship, and an adapted distribution were built patiently and often "undetected." Today, that capability constitutes a key competitive edge over the competition, none of whom believed in it. Now it is up to the company to accept this strategy fully and use it to grow the assets that were shaped during its origin.

For more than a century, the car industry has been a prolific source of inspiring management concepts. Recall that Henry Ford was the father of disruptive strategies, revolutionizing the industry with the Model T, the first car for the mass market. The Model T did not owe its success to Ford's crazy idea of selling "a (simple) car painted in any color . . . so long as it's black," but to designing a low-cost product associated with a system of production and distribution, thus opening up automobile mobility to a huge mass of Americans (and others). Following on from our earlier book on the Logan, we hope that this book helps to make clear the lessons to be learned from what we believe is a significant chapter in contemporary industrial history.

Bibliography

Abrahamson, Eric. "Managerial Fads and Fashions: The Diffusion and Rejection of Innovations." *The Academy of Management Review* 16, no. 3 (1991): 586–612.

Anderson, Philip, and Michael L. Tushman. "Technological Discontinuities and Dominant Designs: A Cyclical Model of Technological Change." *Administrative Science Quarterly* 35, no. 4 (1990): 604–635.

Barlatier, Pierre-Jean, and Olivier Dupouët. "Ambidextrie organisationnelle et structure de la firme : une approche dynamique." In *Proceedings Congrès AIMS*, Grenoble. June, 2009.

Bartlett, Christopher A., and Sumantra Ghoshal. "Managing Innovation in the Trans-National Corporation." In *Managing the Global Firm,* Christopher A. Bartlett, Yves L. Doz, and Gunnar Hedlund (eds.). London: Routledge, 2011, pp. 215–255.

Beaume, Romain. *L'Ingénierie Avancée & les Programmes Pilotes ; Les Dynamiques d'Innovation Automobiles.* Thèse de l'Ecole Polytechnique, Paris, 2012.

Beise, Marian. *Lead Markets: Country-Specific Drivers of the Global Diffusion of Innovations.* Heidelberg: Springer, 2001.

Beise, Marian. "Lead Markets: Country-Specific Drivers of the Global Diffusion of Innovations." *Research Policy* 33, no. 6–7 (2004): 997–1018.

Beise, Marian. "Innovation Units Within Established Firms. Towards a Cartography." *IPDMC Proceedings*. Copenhagen. 2015.

Ben Mahmoud-Jouini, Sihem, Florence Charue-Duboc, and François Fourcade. "Favoriser l'innovation radicale dans une entreprise multidivisionnelle : Extension du modèle ambidextre à partir de l'analyse d'un cas." *Finance Contrôle Stratégie* 10, no. 3 (2007): 5–41.

Ben Mahmoud-Jouini, Sihem, and Christophe Midler. "Compétition par l'innovation et dynamique des systèmes de conception dans les entreprises françaises ; une comparaison de trois secteurs." *Entreprises et Histoire,* no. 23 (1999).

Ben Mahmoud-Jouini, Sihem, Florence Charue-Duboc, and Christophe Midler. *Management de l'innovation et Globalisation : Enjeux et pratiques contemporains*. Paris: Dunod, 2015.

Benabdejlil, Nadia, Yannick Lung, and Alain Piveteau. "L'émergence d'un pôle automobile à Tanger (Maroc)." *Cahiers du GREThA*, no. 4 (2016). (To be published in *Critique économique*.)

Benner, Mary J., and Michael L. Tushman. "Exploitation, Exploration, and Process Management: The Productivity Dilemma Revisited." *The Academy of Management Review* 28, no. 2 (2003): 238–256.

Bhargava, R. C. *The Maruti Story: How a Public Sector Company Put India on Wheels*. Noida, India: Collins Business, 2010.

Bonin, Hubert, Yannick Lung, and Steven Tolliday. *Ford 1903–2003: The European History*. Paris: PLAGE, 2003.

Bonnafous, Gilles. *The Renault Technocentre*. Paris: Hazan, 1998.

Bordenave, G., and Yannick Lung. "Twin Strategies of Internationalization of the US Carmakers: GM and Ford." In *Globalization or Regionalization of the American and Asian Car Industry*, Michel Freyssenet, Koichi Shimizu, and Guiseppe Volpato (eds.). London: Palgrave, 2003, pp. 53–94.

Boyer, Robert, Elsie Charron, Ulrich Jergens, and Steven Tolliday, eds. *Between Imitation and Innovation: The Transfer and Hybridization of Productive Models in the International Automobile Industry*. Oxford: Oxford Univ. Press, 1998.

Boyer, Robert, and Michel Freyssenet. *The Productive Models: The Conditions of Profitability*. London: Palgrave, 2002.

Brown, Tim. *Change by Design: How Design Thinking Transforms Organizations and Inspires Innovation*. New York: Harper-Collins, 2009.

Burgelman, Robert A., and Leonard R. Sayles. *Inside Corporate Innovation Strategy, Structure, and Managerial Skills*. New York: Free Press, 1988.

Burgelman, Robert A., Webb McKinney, and Philip E. Meza. *Becoming Hewlett Packard: Why Strategic Leadership Matters*. New York: Oxford University Press, 2017.

Camuffo, Arnaldo, and Guiseppe Volpato. *Global Sourcing in the Automotive Supply Chain: The Case of Fiat Auto "Project 178" World Car*. International Motor Vehicle Programme, MIT Globalization Research, 2002, multigr.

Carrincazeaux, Christophe, and Yannick Lung. "La proximité dans l'organisation de la conception des produits automobiles." In *Proximité(s) : approche interdisciplinaire*, M. Bellet, T. Kirat, and C. Largeron (eds.). Hermès, Paris, 1998, pp. 241–265.

Castoriadis, Cornelius. *Devant la guerre*. Paris: Librairie générale française, 1983.

Chacko, Philip, Christabelle Noronha, and Sujata Agrawal. *Small Wonder: The Making of the Nano.* Chennai: Westland, 2010.

Chandler, Alfred Dupont. *Strategy and Structure: Chapters in the History of the Industrial Enterprise.* Cambridge, MIT Press, 1962.

Charue-Duboc, Florence, and Christophe Midler. "L'activité d'ingénierie et le modèle de projet concourant." *Sociologie du travail* 44, no. 3 (2002): 401–417.

Christensen, Clayton M. *The Innovator's Dilemma: When New Technologies Cause Great Firms to Fail.* Cambridge, MA: Harvard Business School Press, 1997.

Christensen, Clayton M., Michael E. Raynor, and Rory McDonald. "What Is Disruptive Innovation?" *Harvard Business Review* 93, no. 12 (2015): 44–53.

Clark, Kim B., and Takahiro Fujimoto. *Product Development Performance: Strategy, Organization, and Management in the World Auto Industry.* Boston, MA: Harvard Business School Press, 1991.

Dalmasso, Cedric. *Internationalisation des activités d'ingénierie dans l'industrie automobile, les dynamiques d'acteur et de métier dans le processus d'organisation.* Thèse de Doctorat en Sciences de Gestion, Ecole des Mines de Paris, 2009.

Dalmasso, Cedric, and Rémi Maniak. "La genèse d'un centre de R&D à l'international. Le cas de l'industrie automobile." *Management International* 19, no. 4 (2015): 83–94.

Dassbach, Carl H. A. *Global Enterprises and the World Economy: Ford, General Motors, and IBM, the Emergence of the Transnational Enterprise.* New York: Garland, 1989.

Doz, Yves L., and Keeley Wilson. *Managing Global Innovation: Frameworks for Integrating Capabilities Around the World.* Harvard Business Press, 2012.

Dunford, Michael. "Globalization Failures in a Neo-Liberal World: The Case of FIAT Auto in the 1990s." *Geoforum* 40, no. 2 (2009): 145–157.

Fourcade, François, and Christophe Midler. "The Role of 1st Tier Suppliers in Automobile Product Modularisation: The Search for a Coherent Strategy." *International Journal of Automotive Technology and Management* 5, no. 2 (2005): 146–165.

Freiberg, Kevin, Dain Dunston, and Jackie Freiberg. *Nanovation: How a Little Car Can Teach the World to Think Big and Act Bold.* Nashville, TN: Nelson, 2011.

Freyssenet, Michel. "Renault, from Diversified Mass Production to Innovative Flexible Production." In *One Best Way? The Trajectories and Industrial Models of World Automobile Producers.* Michel Freyssenet, Andrew Mair, Koichi Shimizu, and Guiseppe Volpato (eds.). Oxford and New York: Oxford University Press, 1998, pp. 365–394.

Fridenson, Patrick. "Le procès de la R4 n'aura pas lieu." *Entreprises et histoire* 1, no. 78 (2015): 147–149.

Frigant, Vincent, and Bernard Jullien. "Comment la production modulaire transforme l'industrie automobile." *Revue d'économie industrielle*, no. 145 (2014): 11–44.

Frigant, Vincent, and Yannick Lung. "Geographical Proximity and Supplying Relationships in Modular Production." *International Journal of Urban and Regional Research* 26, no. 4 (2002): 742–755.

Fujimoto, Takahiro, and Youngwon Park. "Complexity and Control: Comparative Study of Automobiles and Electronic Products." *MMRC Discussion Paper Series*, no. 352 (2011).

Garel, Gilles. "L'entreprise sur un plateau : un exemple de gestion de projet concourante dans l'industrie automobile." *Gestion 2000*, no. 3 (1996): 111–134.

Garel, Gilles. *Le management de projet*. La Découverte, Repères, Paris, 2011.

Garel, Gilles, Alex Kesseler, and Christophe Midler. "Le co-développement, définitions, enjeux et problèmes. Le cas de l'industrie automobile." *Education Permanente*, no. 131 (1997): 95–110.

Gemünden, Hans Georg. "When Less Is More, and When Less Is Less." *Project Management Journal* 46, no. 3 (2015): 3–9.

Gibson, Christina B., and Julian Birkinshaw. "The Antecedents, Consequences, and Mediating Role of Organizational Ambidexterity." *Academy of Management Journal* 47, no. 2 (2004): 209–226.

Gilbert, Clark G. "Change in the Presence of Residual Fit: Can Competing Frames Co-Exist?" *Organization Science* 17, no. 1 (2005): 150–167.

Goldratt, Eliyahu M., and Jeff Cox. *The Goal: A Process of Ongoing Improvement*, 2nd rev. ed. Croton-on-Hudson, NY: North River Press, 1992. (First edition, 1984).

Govindarajan, Vijay. "Conversations: Reverse Innovation: An Interview with Vijay Govindarajan." *Research-Technology Management* 55, no. 6 (2012).

Govindarajan, Vijay, and Chris Trimble. *Reverse Innovation: Create Far from Home, Win Everywhere*. Boston, MA: Harvard Business Review Press, 2012.

Govindarajan, Vijay, and Ravi Ramamurti. "Reverse Innovation, Emerging Markets, and Global Strategy." *Global Strategy Journal* 1, no. 3–4 (2011): 191–205.

Guerineau, Mathias, Florence Charue-Duboc, and Sihem Ben Mahmoud-Jouini. "Le déploiement des innovations : (re)penser le transfert et la diffusion de l'innovation dans les firmes multinationales." *Communication à la 6ème conférence Atlas–AFMI (Association Française de Management International)*, Nice, 6–8 juin 2016.

Hargadon, Andrew B., and Angelo Fanelli. "Action and Possibility: Reconciling Dual Perspectives of Knowledge in Organizations." *Organization Science* 13, no. 3 (2002): 290–302.

Hatchuel, Armand, and Benoît Weil. "C-K Design Theory: An Advanced Formulation." *Research in Engineering Design* 19, no. 4 (2008): 181–192.

Henderson, Rebecca M., and Kim B. Clark. "Architectural Innovation: The Reconfiguration of Existing Product Technologies and the Failure of Established Firms." *Administrative Science Quarterly* 35, no. 1 (1990).

Hounshell, David A. *From the American System to Mass Production, 1800–1932: The Development of Manufacturing Technology in the United States.* Baltimore and London: The Johns Hopkins University Press, 1985.

Humphrey, John, Yveline Lecler, and Mario Sergio Salerno. *Global Strategies and Local Realities: The Auto Industry in Emerging Markets.* Houndmills, Basingstoke, Hampshire: Macmillan Press, 2000.

Hussler, Caroline, and Thierry Burger-Helmchen. "Inversée vous avez dit inversée? Une typologie stratégique de l'innovation inversée." *Revue française de gestion* 42, no. 255 (2016): 105–119.

Imai, Masaaki. *Kaizen, the Key to Japan's Competitive Success.* New York: McGraw-Hill, 1986.

Immelt, Jeffrey R., Vijay Govindarajan, and Chris Trimble. "How GE Is Disrupting Itself." *Harvard Business Review* 87, no. 10 (2009): 56–65.

Itazaki, Hideshi. *The Prius that Shook the World: How Toyota Developed the World's First Mass-Production Hybrid Vehicle.* Tokyo: Nikkan Kogyo Shimbun, 1999.

Jolivet, Françoise, and Christian Navarre. "Grand projets, auto-organisation, métarègles : vers de nouvelles formes de management des grands projets." *Gestion 2000* 19 (1993): 191–200.

Jullien, Bernard. "Relativiser le statut de la rupture dans la théorie évolutionniste." In *Approches évolutionnistes de la firme et de l'industrie,* Maurice Baslé, Robert Delorme, Jean-Louis Lemoigne, Bernard Paulré (coord.). Paris: L'Harmattan, 1999, pp. 207–228.

Jullien, Bernard, Yannik Lung, and Christophe Midler. *The Logan Epic: New Trajectories for Innovation.* Paris: Dunod, 2013.

Jullien, Bernard, Yannik Lung, and Christophe Midler. "De la Logan à la Kwid. Ambidextrie, innovation inversée et fractale, design-to-cost : les recettes de la stratégie Entry de Renault." *Cahiers du GREThA*, Univ. Bordeaux no. 19 (2016).

Jullien, Bernard, and Tomasso Pardi. "Structuring New Automobile Industries . . . Political Consequences. *ERIEP*, no. 3 (2013).

Khanna, Turan, Krishna G. Palepu, and Jayant Sinha. "Strategies that Fit Emerging Markets." *Harvard Business Review* 83, no. 6 (2005): 63–74.

Kim, W. Chan, and Renée Mauborgne. *Blue Ocean Strategy: How to Create Uncontested Market Space and Make Competition Irrelevant.* Boston, MA: Harvard Business Review Press, 2005.

Laperche, Blandine, and Gilliane Lefebvre. "The Globalization of Research & Development in Industrial Corporations: Toward 'Reverse Innovation'? The Cases of General Electric and Renault." *Journal of Innovation Economics & Management* 2, no. 10 (2012): 53–79.

Laurens, Patricia, and Christian Le Bas. "L'innovation inverse : clarification conceptuelle et essai d'évaluation quantitative." *Mondes en développement* 173, no. 1 (2016): 47.

Layan, Jean-Bernard. "L'innovation péricentrale dans l'industrie automobile : une gestion territoriale du risque de résistance au changement." *Flux,* no. 63–64 (2006): 42–53.

Le Masson, Pascal, Benoît Weil, and Armand Hatchuel. *Strategic Management of Innovation and Design.* New York: Cambridge University Press, 2012.

Lenfle, Sylvain. "Exploration and Project Management." *International Journal of Project Management* 26, no. 5 (2008): 469–478.

Linhart, Robert, *L'Établi,* Paris, Minuit, 1978.

Loubet, Jean-Louis. *Citroën, Peugeot, Renault et les autres. Soixante ans de stratégies.* Paris: Le Monde-Éditions, 1995.

Lundin, Rolf A., Niklas Arvidsson, Tim Brady, Eskil Eksted, Christophe Midler, and Jörg Sydow. *Managing and Working in Project Society: Institutional Challenges of Temporary Organizations.* Cambridge, UK: Cambridge University Press, 2015.

Lung, Yannick. "Repenser les trajectoires de la géographie de l'innovation." In *Les trajectoires de l'innovation.* C. Bouneau and Y. Lung (dir.). Bruxelles, Belgique: Peter Lang, 2014, pp. 201–223.

Lung, Yannick, Jean-Jacques Chanaron, Takahiro Fujimoto, and Daniel Raff, eds. *Coping with Variety: Flexible Productive Systems for Product Variety in the Auto Industry.* Aldershot, Hampshire, England: Ashgate, 1999.

Maniak, Rémi, and Christophe Midler. "Multiproject Lineage Management: Bridging Project Management and Design-Based Innovation Strategy." *International Journal of Project Management* 32, no. 7 (2014): 1146–1156.

Midler, Christophe. "Logique de la mode managériale." *Gérer et Comprendre,* no. 3 (1986): 74–85.

Midler, Christophe. *L'auto qui n'existait pas. Management des projets et transformation de l'entreprise.* InterEditions, Paris, 1993. Nouvelle édition Dunod, Paris, 2012.

Midler, Christophe. "'Projectification' of the Firm: The Renault Case." *Scandinavian Management Journal* 11, no. 4 (1995): 363–375.

Midler, Christophe. "Implementing a Low-End Disruption Strategy Through Multi-project Lineage Management: The Logan Case." *Project Management Journal* 44, no. 5 (2013): 24–35.

Midler, Christophe, Catherine P. Killen, and Alexander Kock. "Project and Innovation Management: Bridging Contemporary Trends in Theory and Practice." *Project Management Journal* 47, no. 2 (2016): 3–7.

Midler, Christophe, and Christian Navarre. "Project Management in the Automotive Industry." In *The Wiley Guide to Managing Projects,* Jeffrey K. Pinto and Peter W. G. Morris (eds.). Hoboken, NJ: John Wiley & Sons, 2004. pp. 1368–1388.

Midler, Christophe, Rémi Maniak, and Romain Beaume. *Réenchanter l'industrie par l'innovation: l'expérience des constructeurs automobiles.* Paris: Dunod, 2012.

Mintzberg, Henry. *The Structuring of Organizations: A Synthesis of the Research.* Englewood Cliffs, NJ: Prentice-Hall, 1979.

Mintzberg, Henry, and Alexandra McHugh. "Strategy Formation in an Adhocracy." *Administrative Science Quarterly* 30, no. 2 (1985): 160.

Mintzberg, Henry, and James A. Waters. "Of Strategies, Deliberate and Emergent." *Strategic Management Journal* 6, no. 3 (1985): 257–272.

Ohmae, Kenichi. *Triad Power. The Coming Shape of Global Competition.* New York: Free Press, 1985.

Pechmann, Felix von, Christophe Midler, Rémi Maniak, and Florence Charue-Duboc. "Managing Systemic and Disruptive Innovation: Lessons from the Renault Zero Emission Initiative." *Industrial and Corporate Change* 24, no. 3 (2015): 677–695.

Prahalad, C. K. *The Fortune at the Bottom of the Pyramid: Eradicating Poverty Through Profits.* Upper Saddle River, NJ: Wharton School Publishing, 2005.

Prahalad, C. K, and Yves L. Doz. *The Multinational Mission: Balancing Local Demands and Global Vision.* New York: Free Press, 1999.

Radjou, Navi, Jaideep C. Prabhu, and Simone Ahuja. *Jugaad Innovation: Think Frugal, Be Flexible, Generate Breakthrough Growth.* San Francisco, CA: Jossey-Bass, 2012.

Radojévic, Nebojša. "Reverse Innovation Reconceptualised: Much Geo-Economic Ado about Primary Market Shift." *Management International* 19, no. 4 (2015): 708–782.

Salerno, Mario S., Mauro Zilbovicius, Ana Valéria Carneiro Dias. "Flexibility Through Modularity: Experimentations of Fractal Production in Europe and Brazil." In *Coping with Variety: Flexible Productive Systems for Product Variety in the Auto Industry,* Yannick Lung, Jean-Jacques Chanaron, Takahiro Fujimoto, and Daniel Raff (eds.). Aldershot, Hampshire, England: Ashgate, 1999, pp. 224–257.

Schweitzer, Louis. *Mes années Renault: Entre Billancourt et le Marché Mondial.* Paris: Gallimard, 2007.

Tiwari, Rajnish, Louise Fischer, and Katherina Kalogerakis. "Frugal Innovation in Scholarly and Social Discourse: An Assessment of Trends and Potential Societal Implications." Joint working paper of Fraunhofer MOEZ Leipzig and Hamburg University of Technology in the BMBF-ITA project, Leipzig and Hamburg, 2016.

Tiwari, Rajnish, and Cornelius Herstatt. *Aiming Big with Small Cars: Emergence of a Lead Market in India*. India Studies in Business and Economics. Springer International Publishing, 2014.

Tushman, Michael L., and Charles A. O'Reilly. "Ambidextrous Organizations: Managing Evolutionary and Revolutionary Change." *California Management Review* 38, no. 4 (1996): 8–29.

Utterback, James M., and William J. Abernathy. "A Dynamic Model of Product and Process Innovation." *Omega* 3, no. 6 (1975): 639–656.

Vernon, Raymond. "International Investment and International Trade in the Product Cycle." *The Quarterly Journal of Economics* 80, no. 2 (1966): 190–207.

Wheelwright, Steven C., and Kim B. Clark. *Revolutionizing Product Development: Quantum Leaps in Speed, Efficiency and Quality*. New York: Free Press, 2009.

Wilkins, Mira, and Frank Ernest Hill. *American Business Abroad: Ford on Six Continents*. Detroit, WI: Wayne State University Press, 1964.

Womack, James T., Daniel T. Jones, and Daniel Roos. *The Machine That Changed the World*. Cambridge, MA: MIT Press.

Zedtwitz, Max von, Simone Corsi, Peder Veng Søberg, and Romeo Frega. "A Typology of Reverse Innovation." *Journal of Product Innovation Management* 32, no. 1 (2014): 12–28.

Zeschky, Marco B., S. Stephan Winterhalter, and Oliver Gassmann. "From Cost to Frugal and Reverse Innovation: Mapping the Field and Implications for Global Competitiveness." *Research Technology Management* 57, no. 4 (2014): 1–8.

Appendix

List of People Interviewed

Carlos Ghosn, Chairman and CEO of Renault-Nissan Alliance

Arnaud Deboeuf, Entry Programme Director from 2012 to 2015; Alliance Senior Vice President, Renault-Nissan BV and the Alliance CEO office from 2015

Gérard Detourbet, Alliance Global Vice President of Renault-Nissan BV and Managing Director of 2ASDU

Design

Patrick Lecharpy, Design France: LCI

Philippe Ponsot, Industrial Design Manager/Chennai

Jean-Philippe Salar, Head of Vehicle Design Studio, Mumbai (has been replaced by Tran-Dhin Sébastien)

Antony Thirion, Industrial Design/Chennai

Sébastien Tran-Dinh, Head of Vehicle Design Studio, Mumbai

Laurens Van Den Acker, Renault's Senior Vice President of Corporate Design

Rnaipl Plant

Colin MacDonald, Plant Director

Jean-Louis Theron, Deputy Managing Director at Renault Nissan Automotive India Pvt Ltd.

Economic Functions

Sylvain Dufeu, Budget Control/2ASDU
Arnaud Hayes, Director of Production Cost, India
Tatjana Todorovic, Production Cost Control/2ASDU

Technical Functions of 2ASDU

Michel Anastassiou, Project Director, Gearbox
Céline Buchard, D-CEV, Developments
Edouard Chainet-Manouvrier, D-CEV, Distribution and Second Internationalisation
Sergio De Carvalho, Department Manager, KWID Body-in-White/2ASDU
Nuno De Morais, Purchasing Director
Philippe Doignon, Department Manager, Electricity/Electronics
Isabelle Ehrart, Department Manager, Gearbox Industrialisation, until 2015; Powertrain Process Engineering Director
Marc-Antoine Gauthier, PPM, Purchasing Project Manager
Christine Genin, Chief Vehicle Engineer, Development of New Vehicles
Ludovic Gouere, Powertrain Product Development Director from 2015
Anash Haime, Management of Suppliers
Kou Kimura, Bodywork Process Engineering Director
Jean-Denis Lenoir, Department Manager, Chassis
Yannick Le Gleut, PPM, Purchasing Project Manager
Fréderic Maniaudet, Customer Requirement Manager
Marc Prin, Project Director, Engine
Thomas Regis, PPM, Powertrain Purchasing until 2015
Jean-François Vial, Chief Vehicle Engineer, Development of KWID

Indian or Regional Functions

Loïc Feuvray, Renault Product Department in Charge of KWID
Patrice Levy-Bencheton, Programme Director for India
Dominique Lucas, Product Department, India

Sumit Sawhney, Renault India's Managing Director
Rafael Treguer, Director, Sales and Marketing, India

Brazil

Bertrand Ciavaldini, Director of 2ASDU Brazil

Index

V

X

9781032476827